Matter: A Very Short Introduction

VERY SHORT INTRODUCTIONS are for anyone wanting a stimulating and accessible way into a new subject. They are written by experts, and have been translated into more than 45 different languages.

The series began in 1995, and now covers a wide variety of topics in every discipline. The VSI library currently contains over 600 volumes—a Very Short Introduction to everything from Psychology and Philosophy of Science to American History and Relativity—and continues to grow in every subject area.

Very Short Introductions available now:

ABOLITIONISM Richard S. Newman
ACCOUNTING Christopher Nobes
ADAM SMITH Christopher J. Berry
ADOLESCENCE Peter K. Smith
ADVERTISING Winston Fletcher
AFRICAN AMERICAN RELIGION
    Eddie S. Glaude Jr
AFRICAN HISTORY John Parker and
    Richard Rathbone
AFRICAN POLITICS Ian Taylor
AFRICAN RELIGIONS
    Jacob K. Olupona
AGEING Nancy A. Pachana
AGNOSTICISM Robin Le Poidevin
AGRICULTURE Paul Brassley and
    Richard Soffe
ALEXANDER THE GREAT
    Hugh Bowden
ALGEBRA Peter M. Higgins
AMERICAN CULTURAL HISTORY
    Eric Avila
AMERICAN HISTORY Paul S. Boyer
AMERICAN IMMIGRATION
    David A. Gerber
AMERICAN LEGAL HISTORY
    G. Edward White
AMERICAN NAVAL HISTORY
    Craig L. Symonds
AMERICAN POLITICAL HISTORY
    Donald Critchlow
AMERICAN POLITICAL PARTIES
    AND ELECTIONS L. Sandy Maisel
AMERICAN POLITICS
    Richard M. Valelly

THE AMERICAN PRESIDENCY
    Charles O. Jones
THE AMERICAN REVOLUTION
    Robert J. Allison
AMERICAN SLAVERY
    Heather Andrea Williams
THE AMERICAN WEST Stephen Aron
AMERICAN WOMEN'S HISTORY
    Susan Ware
ANAESTHESIA Aidan O'Donnell
ANALYTIC PHILOSOPHY
    Michael Beaney
ANARCHISM Colin Ward
ANCIENT ASSYRIA Karen Radner
ANCIENT EGYPT Ian Shaw
ANCIENT EGYPTIAN ART AND
    ARCHITECTURE Christina Riggs
ANCIENT GREECE Paul Cartledge
THE ANCIENT NEAR EAST
    Amanda H. Podany
ANCIENT PHILOSOPHY Julia Annas
ANCIENT WARFARE Harry Sidebottom
ANGELS David Albert Jones
ANGLICANISM Mark Chapman
THE ANGLO-SAXON AGE
    John Blair
ANIMAL BEHAVIOUR
    Tristram D. Wyatt
THE ANIMAL KINGDOM
    Peter Holland
ANIMAL RIGHTS David DeGrazia
THE ANTARCTIC Klaus Dodds
ANTHROPOCENE Erle C. Ellis
ANTISEMITISM Steven Beller

WAVES Mike Goldsmith
WEATHER Storm Dunlop
THE WELFARE STATE David Garland
WILLIAM SHAKESPEARE Stanley Wells
WITCHCRAFT Malcolm Gaskill
WITTGENSTEIN A. C. Grayling
WORK Stephen Fineman

WORLD MUSIC Philip Bohlman
THE WORLD TRADE
    ORGANIZATION Amrita Narlikar
WORLD WAR II Gerhard L. Weinberg
WRITING AND SCRIPT
    Andrew Robinson
ZIONISM Michael Stanislawski

Available soon:

METHODISM William J. Abraham
TOLSTOY Liza Knapp
CONCENTRATION CAMPS Dan Stone

SYNAESTHESIA Julia Simner
READING Belinda Jack

For more information visit our website

www.oup.com/vsi/

Geoff Cottrell

MATTER

A Very Short Introduction

OXFORD
UNIVERSITY PRESS

OXFORD

UNIVERSITY PRESS

Great Clarendon Street, Oxford, OX2 6DP,
United Kingdom

Oxford University Press is a department of the University of Oxford.
It furthers the University's objective of excellence in research, scholarship,
and education by publishing worldwide. Oxford is a registered trade mark of
Oxford University Press in the UK and in certain other countries

Published in the United States of America by Oxford University Press
198 Madison Avenue, New York, NY 10016, United States of America

British Library Cataloguing in Publication Data
Data available

Library of Congress Control Number: 2018962713

ISBN 978-0-19-880654-7

Printed in Great Britain by
Ashford Colour Press Ltd., Gosport, Hampshire.

Contents

Acknowledgements

It is a pleasure to thank my editor, Latha Menon, for suggesting this stimulating topic for a VSI. I would also like to thank Phil Hopkins, Laura Lauro, and Jo Marks, for their comments on the first draft.

List of illustrations

Chapter 1
What is matter?

Matter is the stuff from which you and all the things in the world around you are made. If you had the most powerful microscope imaginable you could look inside your body and see that you are made of atoms. Inside every atom is a tiny nucleus, and orbiting the nucleus is a cloud of electrons. The nucleus is made out of protons and neutrons, and by zooming in on a nuclear particle you would find that inside it there are even smaller particles—quarks. Quarks are the smallest particles that we have seen, and lie at the limit of resolution of the most powerful microscopes of matter. As far as we know, electrons and quarks are not made from anything smaller and so they are called fundamental particles. All matter is made from just these particles.

Atoms are so small that a million of them can fit across the breadth of a human hair. If an apple were magnified up to the size of the Earth, its atoms would be the size of apples. The diameter of an atom is around 10^{-10} m. (When quantities are given as 10 to some power (10^6 say) this is simply 1 followed by 6 zeros, in this case 1,000,000 or one million; those expressed as 10 to some negative power (10^{-6} say) have 1 in the 6th place after the decimal point, that is, 0.000001 or one millionth.) Your body contains roughly 10^{29} atoms; a size of about two metres defines the human scale.

Matter exists in forms of immense variety and complexity. The familiar things around us—books, a table, water, a cat—all have intricate structures and compositions. They are made of vast numbers of atoms, sticking together in clumps of different shapes and sizes. The 'glue' that holds clumps of atoms together and the electrons to the nuclei of atoms is the electrical force of attraction between opposite electrical charges. All the different structures of matter result from the many possible ways in which particles interact to make different physical forms and arrangements in space. The electrical force, in various guises, also produces the various types of interatomic bonds between atoms, joining them together to make molecules, and so it underpins chemistry. Molecules can be as simple as water (two atoms of hydrogen and one of oxygen, or H_2O), or they can be as complex as the millions of atoms in a protein macromolecule in your body.

A substance is an *element* if it cannot be decomposed into two or more different substances by common physical or chemical processes. There are ninety-two naturally occurring different types of chemical elements, and each type has its own unique properties. In the year 1867 only sixty-three of the elements had been discovered. The different atoms were then known to have different *atomic weights* ranging from the lightest, hydrogen, with an atomic weight of 1, up to the heaviest known at the time, lead, with a weight of 207. (The basic chemical unit is the weight of a hydrogen atom, 1.67×10^{-27} kg, which defines the *atomic mass unit*.) At that time the chemists were searching for patterns in their properties that might reveal a deeper structure. The properties of the elements were well known to the Russian chemist Dmitri Mendeleev, who wrote down their names and properties on cards and arranged them in order of their atomic weights. He noticed that the chemical properties had a pattern: they *repeated* at regular intervals, a periodic law. Mendeleev described his discovery: 'I saw in a dream where all elements fell into place as required. Awakening, I immediately wrote it down on a piece of paper, only in one place did a correction later seem necessary.' The pattern showed that were also some elements missing from the table. He left gaps for these, confidently predicting the elements germanium, gallium, and scandium, which were soon discovered.

The modern periodic table (Figure 1) is arranged not by atomic weight, but by *atomic number*, the number of protons in a nucleus, ranging from 1 (hydrogen) up to 92 (uranium). The atomic number is equal to the number of electrons in the atom. Elements heavier than uranium are produced artificially. Henry Moseley developed an X-ray technique to measure the number of protons in the nucleus, and we have him to thank for essentially the modern version of the table. As the atomic number increases along successive horizontal rows the chemical properties repeat in periods of two (hydrogen and helium), then two periods of eight (lithium to neon; sodium to argon), and then three of eighteen. The table is more than a classification scheme; it reveals a pattern that is deeply embedded in nature and the structure of atoms.

The periodic table is governed by quantum laws. The electrons in atoms spread out around the nucleus in what are known as atomic *orbitals*, which form shell-like structures around the nucleus. Atoms seek to minimize their energies, which is what happens when their electron shells are completely full. For example, the first atom with a full shell is helium (atomic number 2), the second is neon (atomic number $2 + 8 = 10$), and the third is argon (atomic number $2 + 8 + 8 = 18$), and so on. These stable filled-shell atoms are the chemically inert *noble gases*, which sit in the last column of the table.

The different atoms are like letters of an alphabet, which combine to make molecules, akin to the words of a language. How many types of molecules are there? The English language has around a quarter of a million words in current use, all based on a twenty-six-letter alphabet. The letter *a* is always an *a*, whether it appears in the word 'cat' or in other words with completely

I	II								

| | | | | | | | | | | | | | | | | | 1
H
hydrogen
1 |

3 Li lithium 7	4 Be beryllium 9
11 Na sodium 23	12 Mg magnesium 24

19 K potassium 39	20 Ca calcium 40	21 Sc scandium 45	22 Ti titanium 48	23 V vanadium 51	24 Cr chromium 52	25 Mn manganese 55	26 Fe iron 56	27 Co cobalt 59
37 Rb rubidium 85	38 Sr strontium 88	39 Y yttrium 89	40 Zr zirconium 91	41 Nb niobium 93	42 Mo molybdenum 96	43 Tc technetium –	44 Ru ruthenium 101	45 Rh rhodium 103
55 Cs caesium 133	56 Ba barium 137	57–71 lanthanoids	72 Hf hafnium 178	73 Ta tantalum 181	74 W tungsten 184	75 Re rhenium 186	76 Os osmium 190	77 Ir iridium 192
87 Fr francium –	88 Ra radium –	89–103 actinoids	104 Rf rutherfordium –	105 Db dubnium –	106 Sg seaborgium –	107 Bh bohrium –	108 Hs hassium –	109 Mt meitnerium –

lanthanoids	57 La lanthanum 139	58 Ce cerium 140	59 Pr praseodymium 141	60 Nd neodymium 144	61 Pm promethium –	62 Sm samarium 150	63 Eu europium 152
actinoids	89 Ac actinium –	90 Th thorium 232	91 Pa protactinium 231	92 U uranium 238	93 Np neptunium –	94 Pu plutonium –	95 Am americium –

1. The periodic table of the elements.

different meanings, like 'bat'. Similarly, each hydrogen atom in a water molecule is identical to those that combine with carbon to form methane (CH_4), a molecule with completely different properties. In theory it is possible to form billions of different stable chemical compounds by combining the elements of the atomic alphabet in different ways.

			III	IV	V	VI	VII	VIII
								2 He helium 4
			5 B boron 11	6 C carbon 12	7 N nitrogen 14	8 O oxygen 16	9 F fluorine 19	10 Ne neon 20
			13 Al aluminium 27	14 Si silicon 28	15 P phosphorus 31	16 S sulfur 32	17 Cl chlorine 35.5	18 Ar argon 40
28 Ni nickel 59	29 Cu copper 64	30 Zn zinc 65	31 Ga gallium 70	32 Ge germanium 73	33 As arsenic 75	34 Se selenium 79	35 Br bromine 80	36 Kr krypton 84
46 Pd palladium 106	47 Ag silver 108	48 Cd cadmium 112	49 In indium 115	50 Sn tin 117	51 Sb antimony 122	52 Te tellurium 128	53 I iodine 127	54 Xe xenon 131
78 Pt platinum 195	79 Au gold 197	80 Hg mercury 201	81 Tl thallium 204	82 Pb lead 207	83 Bi bismuth 209	84 Po polonium –	85 At astatine –	86 Rn radon –
110 Ds darmstadtium –	111 Rg roentgenium –	112 Cn copernicium –		114 Fl flerovium –		116 Lv livermorium –		

64 Gd gadolinium 157	65 Tb terbium 159	66 Dy dysprosium 163	67 Ho holmium 165	68 Er erbium 167	69 Tm thulium 169	70 Yb ytterbium 173	71 Lu lutetium 175
96 Cm curium –	97 Bk berkelium –	98 Cf californium –	99 Es einsteinium –	100 Fm fermium –	101 Md mendelevium –	102 No nobelium –	103 Lr lawrencium –

The scale of matter

To fix ideas it is important to appreciate the vast range of the
length scales of matter, indicated in Figure 2. From the very
smallest structures that we believe may exist, to the largest (the
visible universe), we cover a mind-bending range of *sixty-two*

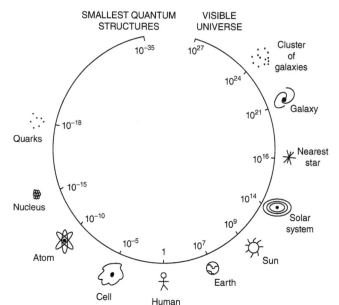

Matter

2. The different length scales of matter (in metres) arranged on a circle.

orders of magnitude in size, or 10^{62}. This takes us from the quantum world of the smallest entities, to that of the largest structures, which are dominated by the force of gravity. The human length scale is at the bottom of the circle, lying roughly in the middle between the two arms of the diagram. The largest structures range from objects the size of the Earth, up to the clusters and superclusters of galaxies. Albert Einstein's *general theory of relativity* describes these huge structures and the space that they occupy. At the very small end of the size spectrum, the laws of *quantum mechanics* describe matter. At present there is no complete theory connecting the quantum world and the gravitationally dominated world, and so the various theories of quantum gravity that have been proposed, which might bridge the gap, are not described here.

Let's take an imaginary zoom lens and, starting from the human scale, zoom in to progressively smaller scales. To see the cells in your body, you would need to increase the magnification 100,000 times from the human scale. This is possible using the wavelengths of visible light. However, light cannot be used to resolve structures smaller than its wavelength. The wavelengths of light are around 500 nanometres (one nanometre, or nm, is 10^{-9} metres, and there are 20,000 wavelengths in a centimetre), and to see smaller things we must use shorter wavelengths. Electrons accelerated in electron microscopes have wavelengths small enough to enable us to see structures as small as an atom. In zooming in from cells to atoms, the magnification has to be increased by 100,000 times. A further increase of magnification of 100,000 times fills our frame with the atomic nucleus. With a total magnification now of a trillion times, we have arrived at the scale of the quantum world, where the wave nature of matter makes things appear very fuzzy. To zoom in beyond this, and look inside the proton, we must accelerate electrons to high-energies and speeds of over 99.9 per cent of the speed of light and crash them into nuclei to see the substructures inside the proton, the quarks. (The speed of light in vacuum is $c = 300,000$ kilometres per second.) There are no microscopes to take us any further on this journey, and from now on one must rely on theory. The finest divisions of space that are believed to exist and have any meaning in terms of the laws of physics are tiny quantum fluctuations on a scale of 10^{-35} metres. An enormous leap in magnification of 10^{17} would be needed to see any structures on this scale.

If we start again on the human scale and zoom *out* ten million times, the field of view is filled by the Earth. From now on, big structures are shaped by gravity. A further zoom of 100 times fills the frame with our star, the Sun. The next largest structure, the solar system, has a diameter of 300 trillion metres, and, to bring that into view, we must zoom out 10,000 times further still. It takes light around five hours to reach Pluto in the outer solar system. We must wait 4.2 years for the light which is now leaving

7

the surface of the next nearest star to the Sun, *Proxima Centauri*, to reach us—the star is 4.2 light years away. As we continue to zoom out an awe-inspiring sight comes into view: our home galaxy the *Milky Way*. The Milky Way is a disc-shaped spiral galaxy containing 100 billion stars and has a diameter of 100,000 light years. To fit this in the frame of the lens we would have to zoom out 10,000 times more.

But the Milky Way is not, as was believed only a century ago, the entire universe. Our galaxy is one of about thirty members of the *local group* of galaxies; to fit the local group in our frame we would need to zoom out around ten times more. Beyond this, there are even larger structures, such as the Coma cluster of galaxies, which has a diameter of over 300 million light years and contains 1,000 galaxies, gravitationally bound together into a roughly spherical clump. To see this cluster, we would need to zoom out a further 100 times. The largest known structures in the universe are the giant *superclusters of galaxies*, and the huge filaments of galaxies which surround vast voids in space, but beyond that, the furthest distance that we can see out to with telescopes is the boundary of the *visible universe*, which has a diameter of around 100 billion light years. All observable matter is contained in this sphere, consisting of 100 billion galaxies, with a total matter content of 10^{80} hydrogen atoms. We will look at these very large scales in Chapter 9, but it's worth noting that the average density of matter in the visible universe is about a few hydrogen atoms per cubic metre. For comparison, the Earth has a density almost 10^{30} times larger than that, making our planet a highly atypical region of the universe. Most of the universe is empty space.

In this book we will, in Chapter 3, see how the familiar states of matter of solid, liquid, and gas arise, and look at some other states of matter. If, like Isaac Newton, we consider matter to be defined as *mass*, then the equivalence of mass and energy, described in Chapter 4, takes our understanding of matter to a deeper level, and reveals the origin of the awesome release of nuclear energy from the atomic nucleus. In Chapter 5 we will enter the weird and fuzzy quantum world and see how it explains the structure of atoms and the periodic table. When large numbers of particles aggregate together, they can, under certain conditions, manifest dramatic and coherent quantum behaviour on macroscopic scales. This is described in Chapter 6, and has led to the development of quantum measurement devices that allow the basic unit of mass, the kilogram, to be defined in terms of the fundamental constants of nature to an unprecedented degree of precision. The ultimate building blocks of matter, which include antimatter, are introduced in Chapter 7, which describes how the world can be understood in terms of around twenty different quantum fields. Most of the mass of normal matter can be explained by the energy in these quantum fields. To understand where the elements come from, in Chapter 8 we look at the history of the universe, from its earliest moments, and trace how the elements are built in stars as well as in the most violent events in the universe. Matter is energy and energy curves space. This property is being put to use by astronomers to map out how much matter the universe contains, and where it is located. Finally, in Chapter 9, we arrive at the humbling realization that the normal matter, the atoms and molecules of familiar everyday life, represent only 5 per cent of all the types of matter that is 'out there'. The remainder of the matter in the universe seems to consist of two completely mysterious substances: dark energy, and dark matter. First, however, we need to assure ourselves that the basic constituents of matter, atoms, really do exist. This is the subject of Chapter 2.

Chapter 2
Atoms

The birth of Western science can be traced back to the Greek philosophers of antiquity. In around 500 BC Thales of Miletus founded a school that sought to explain the world by applying logic and reason to the observation of nature. He proposed that all matter was composed of a single primary substance, which he believed to be water. What was important was not that he was mistaken about this, but that he set in motion a way of thinking that was based in looking at the world as it is *in itself*. This approach led to the prescient idea that the building blocks of matter are *atoms*, a conjecture that is attributed to Leucippus and Democritus in about 450 BC.

Take an apple and chop it up into smaller and smaller pieces. Eventually a point is reached where no more chopping is possible, revealing the ultimate graininess of matter, atoms. The word atom comes from the Greek word *atomos*, meaning indivisible. These ultimate particles were considered to be indestructible, differing only in shape, size, position, and arrangement. The Greeks imagined a space in which the atoms move around ceaselessly, called *void*. The atomist's beautifully simple view was that the universe is composed of just two elements: atoms and void.

In about 350 BC, Aristotle adopted the belief that various combinations of just four elements, earth, fire, air, and water, can explain everyday matter. So great was Aristotle's philosophical standing that this view persisted unchallenged into the medieval period, and inspired the alchemists' fruitless search for ways of turning base matter such as lead into gold. In their unquestioning adherence to the belief in a world made of four elements, the closest the alchemists came to an effective scientific method was a basic form of empirical chemistry. This involved grinding, mixing, heating, and distilling common substances such as water, oil, mercury, earth, sulphur, salt, and air. The dark ages of Europe lasted from the 6th to the 14th century, marking a period when the development of science largely ground to a halt.

Writing at the start of the 17th century, the French philosopher René Descartes considered the primary quality of matter to be *extension*, namely that which occupies length, breadth, and height in space, or as we would say, *volume*. This period marked the birth of the age of enlightenment when quantitative science really began to take off. From about that time the atomic hypothesis began to re-emerge as needed by theories of the day, but there was still no proof of the existence of atoms.

The property of *mass* is central to the concept of matter. Mass first appeared in Newton's laws of motion, which form the basis of modern classical mechanics. Newton asked if there is any simple rule by which the movements of the planets can be calculated given their states of motion. In this quest he developed the mathematical calculus necessary to solve equations describing the change in the state of motion of a mass point in an infinitesimal time, under the influence of an external force. He connected the concept of a force, already well known from the study of statics, with acceleration, and in doing so introduced the concept of mass. The meaning of mass in Newton's second law (*force = mass × acceleration*) is strictly *inertial mass*, the resistance of a body to a change in its motion.

Newton's great breakthrough was to link the laws of motion with the law of gravitational attraction. The force on a mass is determined by the positions and masses of nearby bodies. Newton's *law of gravitation* states that the gravitational force between two masses is equal to a constant (Newton's gravitational constant, G) multiplied by each mass and the inverse square of their separation. Gravity is a universal long-range force (namely, all masses attract each other). The law of motion combined with the law of attraction enabled him to calculate the past and future states of masses acting under the force of gravity. With this he succeeded in explaining the motion of the Moon, the planets, comets, and even the tides with great precision.

What was Newton's view of matter? In his *Optiks* of 1704, Newton writes of matter as being formed of '*solid, massy, impenetrable, moveable particles... so very hard, as never to wear or break in pieces...*', a conception of atoms not very different from that of Democritus. What was new was the deterministic manner in which the 'massy particles' move under the action of forces.

Weight and mass are easily confused, but there is an important difference between them. When we weigh an object what we are really doing is measuring how strongly it is pulled by gravity. If we load a spring balance with a kilogram of apples, a big mass (the Earth) pulls the apples (a small mass) downwards against the pull of the spring. By symmetry, the small mass also pulls the big mass; the force attracting them together is equal and opposite, whether we think of the Earth pulling the apples, or the apples pulling the Earth. This expresses Newton's third law. However, the Sun exerts a gravitational pull on the Earth that is an enormous 10^{22} times stronger than that of the apples, and so the apples have a negligible effect on the Earth. This is why the Earth's orbit is controlled by the Sun and not by apples. (By symmetry, the Earth also exerts a force on the Sun. However, because the mass of the Earth is only three millionths of the mass of the Sun, the effect is small. The two bodies orbit around their common centre of mass, which lies well within the body of the Sun.) If we take our kilogram of apples to the Moon, which has only one-sixth of the mass of the Earth, the balance registers only a sixth of a kilogram. Take them out in space, far from any gravitating matter or into orbit around the Earth on a space station and they exist in a state of free-fall and are *weightless*. Mass is therefore an intrinsic property of an object; whereas the weight of the same object would be different if the object were on the Earth, or the Moon, or on Mars.

After Newton's brilliant achievements, the story of matter and atoms shifts to the work of the early chemists. By the end of the 18th century, important advances had been made in quantitative chemistry, largely attributable to the development of accurate balances to measure the weights of reacting substances. The language chemists used also made more frequent use of the word 'element', of which there were then known to be about thirty. The first quantitative chemical experiments of Joseph Priestley, the discoverer of oxygen, and Antoine Lavoisier showed that familiar substances were often combinations of elements such as hydrogen, oxygen, carbon, iron, and sulphur. In 1789 Lavoisier discovered the law of *conservation of mass*; namely, when substances combined, the mass of the reacting chemicals was always equal to the mass of the products. For example, the mass of the hydrogen plus the mass of the oxygen is always equal to the mass of the water produced. A tragic fate befell Lavoisier. He was a tax collector and an aristocrat who lived at the time of the French Revolution; he was branded a traitor and guillotined in 1794. The mathematician Joseph Lagrange paid tribute to him: 'it took them only an instant to cut off his head, and one hundred years might not suffice to reproduce its like.'

Atomic theory gained a stronger foothold in the early 19th century with the experiments of the English chemist John Dalton. Dalton proposed that each chemical element is a unique type of atom, differing from others by its weight. Dalton discovered simple numerical ratios in the proportions of the weights of elements when they combined chemically. Dalton's law of *constant proportions* states that when elements form compounds, their weights always combine in simple whole-number ratios. For example, the weight of oxygen required to combine with a given weight of hydrogen to form water is always eight times as large. Dalton made an ordered table of the relative weights of the elements: hydrogen, nitrogen, carbon, oxygen, phosphorus, sulphur, copper, lead, silver, gold, platinum, and mercury. This table was an inspiration to Mendeleev and others, and was a precursor to the modern periodic table.

The simple number ratios coming out of chemistry led many to believe in the existence of atoms, but others still harboured doubts. Chemists had revealed many facts about atoms and molecules, but were unable to measure their sizes and absolute weights. The key observation that would ultimately lead to the certainty of the existence of atoms was, it seems, first made in antiquity. In the 1st century BC, the Roman poet Lucretius had followed in the footsteps of the atomist Democritus when he wrote his great epic poem *De Rerum Natura* (On the Nature of Things). It is fortunate that Lucretius was aware of the earlier work, because most of Democritus' original books and manuscripts were destroyed in the great fire that destroyed the library at Alexandria in Egypt in about 48 BC. In his poem, Lucretius describes a darkened room, pierced by a shaft of sunlight entering through a hole in a shutter. The brilliant light illuminates myriads of minute dust particles caught in the beam. They tremble agitatedly, seemingly being jostled at random. Lucretius suggests that this motion results from hordes of unseen atoms, raining repeated tiny blows upon the dust particles.

The same phenomenon was seen in 1827 by the English botanist Robert Brown who had been studying pollen grains in water. The tiny particles danced about (Figure 3), as if they had a life of their own. At the time there was much interest in *vitalism*—the search for a hypothetical life force, which was thought to infuse living matter. Brown wondered if he had discovered the vital force,

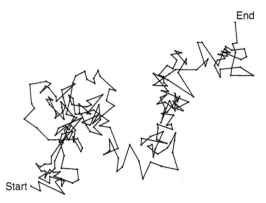

End

Start

3. Brownian motion: the track of a pollen grain particle in water. The grain is gradually nudged away from its starting position as it is jostled by molecules and performs a 'random walk'.

and performed the crucial experiment. He replaced the pollen grains by particles of finely powdered silica, but the particles kept on jiggling. *Brownian motion*, as it has come to be called, was not the vital force.

It took the genius of Einstein to see what Brownian motion was saying, and he used the laws of chance to deduce the existence of molecules, and their sizes. Even if atoms are themselves too small to be seen, they are able to produce a sensible motion in other very small particles that *are* visible. A micron-sized (10^{-6} metres) dust particle in air can be thought of as a super-large air molecule that receives around 10,000 molecular blows on one surface at any instant. A similar number of blows batter the opposite side, so that on average there is no net force on the particle. However, the number of blows is subject to statistical variations. The laws of probability tell us that the number of molecular blows fluctuates as the square root of the number itself. The number of blows received by a dust particle varies between 9,900 and 10,100, making the force on it fluctuate by about 1 per cent, which accounted for the 'random walk' of Brownian motion. Einstein's paper was published in his *annus mirabilis* year of 1905 and it contained the first ever estimate of the size of a molecule.

It is well known that oil and water don't mix. Oil spreads out into a thin layer two molecules thick on the water with the oil molecules lined up side by side, and back to back. The polymath Benjamin Franklin was interested in the effect of pouring oil on 'troubled' waters and the effect it is reputed to have in calming them down. In 1774 he poured a teaspoon of oil on the surface of a pond in Clapham Common in London and, as it spread over a great area, he noted that the surface became 'as smooth as a looking glass'. The English physicist Lord Rayleigh later used this effect in an elegant and simple experiment to measure the size of molecules. The area covered by the oil is easier to see if the water is first lightly dusted with fine powder, which the oil will push away when it spreads out. A single drop of oil with a volume of one cubic millimetre covers an area of about a square metre. The molecular size (about two nanometres) is simply the droplet volume divided by the area of the film. Since there are approximately 12 atoms in an oil molecule, the size of an atom works out to be about 1.7×10^{-10} metres.

So far, we have seen how the prescient ideas of the Greek atomists 2,500 years ago were ignored for centuries until re-emerging at the start of the age of the enlightenment. By the time of Newton, there were three ideas for a definition of matter: all matter is made of atoms, atoms take up space, and atoms have mass. The concept of atoms becomes more firmly rooted by the time of Dalton, and his discovery of the simple ratios of atomic masses in chemical reactions. But it was not until Einstein's interpretation of Brownian motion in 1905 that any objections to the atomic hypothesis were finally quashed.

It is now possible to image individual atoms. In Figure 4, a technique called *Scanning Tunnelling Microscopy* (STM) was

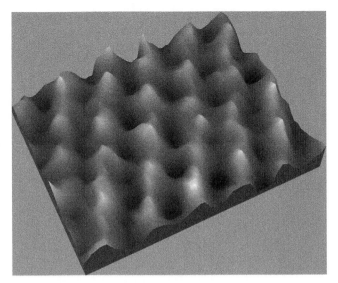

4. Scanning Tunnelling Microscopy (STM) image of the electron clouds of individual carbon atoms on a graphite surface.

used to image individual carbon atoms sitting on the surface of a sample of graphite. The method relies on the ability of electrons to 'tunnel', quantum mechanically, across the gap between the sample and a fine-tipped metal probe. Quantum mechanical tunnelling is described in Chapter 5.

The atomic hypothesis is so fundamental that in 1970 it prompted American physicist Richard Feynman to write in Volume 1 of his famous *Lectures on Physics*:

> If, in some cataclysm, all of scientific knowledge were to be
> destroyed, and only one sentence passed on to the next generations
> of creatures, what statement would contain the most information in
> the fewest words? I believe it is the atomic hypothesis (or the
> atomic fact, or whatever you wish to call it) that all things are made

of atoms—little particles that move around in perpetual motion, attracting each other when they are a little distance apart, but repelling upon being squeezed into one another.

In Chapter 3 I take up the story of how attractive and repulsive forces between atoms produce the familiar states of matter.

Chapter 3
Forms of matter

Water is one of the few everyday substances that can exist naturally on the Earth as solid, liquid, and gas. Cool it down, and it turns into ice as hard as rock. On Saturn's moon Titan, the temperature is a chilly 180 °C below zero and there are mountains of ice, 3 kilometres tall. On our more temperate planet the normal state of water is liquid. When you boil a kettle a jet of invisible gas, steam, is produced. The white clouds that come out of the kettle contain minute drops of water that condense in the air and scatter light. Place a cold surface in the jet and the steam condenses back to drops of water that run down and coalesce. These different states or phases of matter arise because of a competition between opposites: the thermal motion driving particles apart and attractive interatomic forces pulling them together, repulsion and attraction. The 'glue' that holds electrons to atoms, brings atoms together to form molecules, and draws molecules together to make solids and liquids, is electricity. Electrical forces lie behind chemistry, biology, and life itself. States of matter that can flow, the liquids and gases, are called fluids. Solids, liquids, and gases are the so-called *great states of matter*. A solid has a shape and a volume, a liquid has a volume but no shape, and a gas has neither shape nor volume. Liquids and solids are called the condensed states of matter.

The simplest state of matter is gas. In the nineteenth century James Clerk Maxwell and Ludwig Boltzmann developed the kinetic theory of gases that forged a link between the statistical microscopic world of molecules and the macroscopic properties of gas. An 'ideal' gas consists of a large number of atoms or molecules flying about randomly and colliding as if they are perfectly elastic miniature billiard balls. Between their brief collisions they move in straight lines, which is why gases fill containers of any shape and size. There are a vast number (of the order of 10^{22}) of molecules in a litre of gas at standard conditions, a number so large that we can calculate their statistical behaviour with great certainty. When a molecule strikes the wall of its container it imparts a tiny impulse to it and bounces back. The relentless battering of the Lilliputian blows of the molecules averages out to produce a sizeable macroscopic force, the *pressure*, which pushes equally on all the walls.

Even though there are so many molecules in a gas, they are very small, and there is a great deal of space between them. This property of gases makes them *compressible*. If you liquefy air by strong cooling, it contracts to 1/2000th of its volume. The walls of a gas container can be squeezed to compress the gas, like the piston in a bicycle pump. The 17th-century English scientist Robert Boyle performed experiments on gases, which he described elegantly as 'touching the spring of the air'. Boyle discovered that the gas pressure increases in proportion to the inverse of its volume. If one halves the volume of the container, the molecules, now squashed into half the space, bombard the walls twice as intensely, so doubling the pressure.

The molecules in a gas do not all travel at the same speed. The slower ones gain kinetic energy because faster ones strike them more often, and the faster ones lose energy more frequently by colliding with slower ones. Collisions lead to a statistical distribution of particle speeds called the Maxwell–Boltzmann distribution, in which the average speed is related to the *temperature* of the gas. An air molecule at room conditions has an average speed of around 350 m/s, which is roughly the speed of sound, and sound propagates as waves of compressions and rarefactions of the air. The link between temperature and molecular speed implies that there must be a lowest possible temperature, *absolute zero*, at which motion stops. Absolute zero is –273 °C, or zero degrees on the absolute temperature scale measured in Kelvin (0 K).

There is an important distinction between heat and temperature. Some bodies have high temperatures but contain little heat; others are cooler and contain a great deal of heat. Heat depends on *both* the temperature *and* how many particles are involved. Two equal pans of boiling water contain twice as much heat as one, even though they both have the same temperature (100 °C).

Thermodynamics

The powerhouse of the industrial revolution was the steam engine. In trying to improve the performance of steam engines, the transformations between heat, work, and energy were studied intensively and the knowledge gained grew into the science of thermodynamics. There are two basic forms of energy: the energy of motion (or *kinetic energy*), and the energy that the body has by virtue of its position in a force field (or *potential energy*). Lifting a 1 kg mass through a height of a metre in the Earth's gravitational field increases its potential energy by almost 10 Joules. The Joule (J) is the SI unit of energy, a conveniently sized unit for macroscopic bodies. It takes about 400,000 Joules to boil a kilogram of water, which is enough energy to lift a person a height of 500 metres.

An object can gain kinetic energy by increasing its speed. But when the object is taken to the top of a building, it gains potential energy from the gravitational field of the Earth. If it is now dropped, the object picks up speed as it falls, converting its potential energy to kinetic energy. Energy is a useful concept because of the principle of *the conservation of energy*: the sum of the potential and kinetic energy is constant throughout the motion of the object. When the object hits the ground, its energy is converted into the random motions of molecules, or heat.

The fact that heat is a form of energy meant that it had to be included in the law of the conservation of energy, which became the *First Law of Thermodynamics*. When a body (called a 'system' in thermodynamics) is in thermodynamic equilibrium, the system can be thought of as being in a sealed box from which no heat energy can escape or be added. The energy of a thermodynamic system is conserved.

It is important to distinguish between useful forms of energy, such as the interchangeable kinetic or potential energies, and energy that has become so degraded that it can't be used to perform work. It's easy to degrade energy into a useless form, and much harder to reverse the process. The kinetic energy of a moving car, the directed energies of all its components, will, on braking, largely end up as heat. The ordered motion of the car has now become disorganized motion, the random motion of the atoms and molecules in the brake drums. That heat energy is no longer useful in making the car move again. The irreversible nature of heat is embodied in the *Second Law of Thermodynamics*. The first law says that we cannot get something for nothing, and the second law says that we can't even break even! The *quality* of energy, or its ability to do useful work, is related to the amount of disorder in a system, and is measured by thermodynamic quantity called the *entropy*. Entropy was first defined by Rudolf Clausius and put on a statistical molecular basis by Boltzmann. The concept of disorder in a system of particles plays a fundamental role in determining how various arrangements of atoms give rise to the different states of matter.

Each particle in a system has a number of *degrees of freedom*, or independent ways in which it can move or absorb energy. For a gas in equilibrium, the total energy is distributed equally between all the degrees of freedom of the particles, according to the principle of the *equipartition of energy*. A simple monatomic gas, like neon, has three degrees of freedom, which correspond to the three dimensions of space. Molecules have extra degrees of freedom. The chemical bond in a diatomic H_2 molecule for example behaves like a spring connecting two atomic masses that can vibrate or rotate, and it is these that bring in extra degrees of freedom. When a standard amount of matter in a given state increases its temperature by one degree, it absorbs a quantity of heat called the *specific heat*.

Solids and liquids

What happens when we cool a container of gas? The gas molecules rebound from the walls of the container with reduced energies and the temperature of the gas starts to fall. Moving more sluggishly, the molecules are now less independent, and spend more time near their neighbours where interatomic forces slow them down. Interatomic forces result from the electrostatic attraction of atoms or molecules at fairly short range (of a few atomic diameters) but become repulsive at very close spacing. Think of the force between two atoms. At large distances (say more than 10 atomic diameters), there is virtually no force because the positive and negative charges in each atom cancel out almost completely. But when the atoms approach each other, they reveal their granular nature and start to feel each other's internal structures, each being closer to some of the other's atomic charges than others. The negative electron clouds of one, and the positive nuclei of the other, attract. However, if the atoms get too close their outer electron clouds overlap and repulsion sets in.

As the temperature of the gas falls, the balance between the disordered thermal motion and the attractive interatomic forces now swings in favour of the latter as the gas condenses into a *liquid* (Figure 5). The molecules are now close enough to resist

Solid   Liquid   Gas   Plasma

5. Transformations of the states of matter with increasing temperature from solid to liquid, to gas, and to plasma.

bulk compression, and this is the reason why liquids are largely incompressible and are used to transmit forces through pipes in hydraulic systems. At the molecular level, a compromise is struck between attraction and repulsion, which results in the molecules having typical equilibrium spacings of about 3×10^{-10} metres. The molecules of a liquid have just enough thermal energy to enable them to swap places with their neighbours by sliding around them, giving a liquid the fluid property of being able to adapt to the shape of its container.

When the liquid is cooled still further, thermal motion becomes more feeble. The liquid freezes to a *solid*, and all fluidity disappears. The molecules, while still continuing to vibrate weakly, become 'locked' into specific locations and the material becomes hard and takes on a definite shape, with each molecule having a definite position in the solid. The molecules in a crystalline solid occupy specific positions on a regular periodic 3D lattice.

A fundamental difference between a solid and a liquid is the degree to which the molecules maintain their regular ordering patterns over large distances. The hallmark of a crystalline solid is the presence of *long-range order*, where the regular periodicity of the molecular arrangement extends over many lattice spacings. A liquid on the other hand is isotropic and homogeneous. It has the disorganized structure of a gas but its molecules clump together under internal forces, without external pressure needing to be applied. This is possible because there is a critical temperature, above which some external pressure is needed to help the molecules stay close together and oppose thermal motion. As the critical temperature is approached, the gas and liquid phases merge, with a smooth transition between them. The liquid has the disordered structure of a gas, but differs from it by being able to maintain a stable volume without the need of an external container. In passing through a phase change, a system of molecules must break their bonds and reform them in new ways. A quantity of energy, the *latent heat*, has to be provided to do this, which does not contribute to the kinetic energy of the particles. While these microscopic structural changes are taking place, the temperature of the material stays constant, despite the addition or removal of heat energy.

Symmetry

The regular arrangements of atoms and molecules on crystal lattices reveal a deep aspect of matter: *symmetry*. We are very familiar with the ideas of symmetry, for example the symmetries in geometric patterns such as repeating wallpaper or tiling patterns in 2D; also those of a perfect 3D sphere. A sphere can be rotated through any angle, around an infinite number of possible axes or be mirror-reflected about an infinite number of planes passing through its centre, *and it still looks the same.*

A symmetry operation is defined as an action that can be performed on an object that leaves the object unchanged. For example, a crystal lattice can be shifted by a whole number of lattice spacings along one of its lattice directions and it looks identical. A crystal has *discrete symmetries* such as translational symmetry, and also a set of rotational and reflectional symmetries. However, in the disordered liquid and gas phases there are an infinite number of *continuous* symmetry operations. Matter in these phases can be translated, reflected, and rotated in an infinite number of ways and it still looks the same. When matter is condensed from the high-energy disordered gaseous or liquid phases into a solid crystal, the degree of symmetry is reduced, giving rise to what is called *broken symmetry*.

A classic example of symmetry breaking occurs in magnets. A permanent magnet, such as a fridge magnet, is composed of an array of microscopic magnets. When the internal magnets all line up and point in the same direction their fields add to produce a strong overall magnetic field, and the body as a whole behaves like one single magnet. This configuration has a lower energy than the very many possible states in which the internal magnets all point in random directions. However, if the magnet is heated, there is a critical temperature, the *Curie temperature*, above which all magnetism is lost. Above this temperature, the random thermal motion becomes so strong that the internal magnets can no longer stay lined up, and instead point in all possible directions. This high temperature state is one of maximum disorder, or entropy, and causes the magnet to lose its magnetism. On cooling, however, the whole assembly spontaneously 'flips' back, to its lower energy state, with the internal magnets again pointing in one preferred direction. The final direction can be anywhere. However, if the material is already threaded by a weak 'seed' magnetic field, such as the Earth's field, they can align to that. This sudden onset of magnetism is a classic example of *spontaneous symmetry breaking*, and marks the transition from a high temperature, high entropy state, to a low temperature, low entropy one. The energy landscape of this process is illustrated in Figure 6 in the example of the unstable equilibrium of a ball balanced on the top of a perfectly symmetric hill. When the ball rolls down a valley it lowers its potential energy and the symmetry is broken.

Symmetry is built into the laws of nature at the most fundamental level. Every continuous symmetry found in nature is associated with a conserved or invariant quantity, as was proved in a famous theorem by Emmy Nöther in 1915. The properties of space and time tell us that the laws of nature are the same everywhere in the

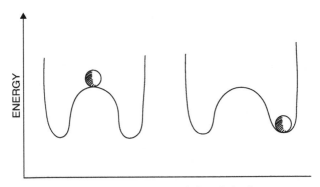

6. Illustration of an unstable symmetry (left) and a broken symmetry (right).

universe. For example, the law of the conservation of linear momentum is independent of the choice of the origin of the coordinate system in which the motion of bodies is measured. Space has a symmetry called *translational invariance*. Also, the fact that the laws of physics are the same at all times turns out to be equivalent to the law of the conservation of energy. The conservation of energy does not depend on what time a clock is set to. Space is also isotropic; the law of the conservation of angular momentum does not depend on where the axis of a spinning body points in space.

Sticking together

Interatomic forces are all different manifestations of the electrical force; in 1785, Charles-Augustin de Coulomb discovered the underlying law. The 'Coulomb' force between two charged bodies is proportional to the product of their charges and, like the gravitational force, is a long-range force, being inversely proportional to the square of their separation. However, the electrical force is enormously stronger than gravity; the repulsive electrical force between two electrons is a factor of 10^{42} stronger than the gravitational force which attracts them. Given this huge difference, why do we mainly experience gravity and not electricity on the human scale? Macroscopic objects are made from enormous numbers of positive and negative charges, which are so intimately mixed that the combined forces of attraction of unlike charges and repulsion of like charges cancel each other out almost perfectly. Lumps of matter are therefore almost completely electrically neutral.

Occasionally a small amount of charge can be transferred by friction from one body to another, creating an imbalance. This is *static electricity*, well known since antiquity, where actions like pulling off a sweater or combing hair make sparks jump or hair stand on end. In 585 BC Thales describes how a piece of amber (which is fossilized tree resin), when rubbed on fur, can pick up lightweight objects like feathers. The Greek word for amber is *elektron*, from which we get our word for electricity.

The first fundamental particle of matter to be identified was discovered in 1897 by J. J. Thompson when he chipped electrons off atoms in electrical discharges. His apparatus, a 'cathode ray tube', was an evacuated glass tube with two sealed-in electrodes. When a high voltage is applied to the electrodes, streams of electrons ('cathode rays') are projected in straight lines from the negative electrode (the cathode), through the tube towards the positive electrode (the anode). Wherever the 'rays' strike the glass, it glows with a mysterious yellowish-green fluorescence. By bringing up electrically charged plates and magnets, the cathode rays can be deflected in a systematic way, and Thompson used these fields to measure the electric charge (e) and mass of the electron. His discovery that atoms contain smaller particles, electrons, ended forever the concept of the indivisible Greek atom.

Chemical bonds

We picture atoms as having a positive nucleus, surrounded by a cloud of electrons flying about. All atoms are attracted to each other, by weak forces arising from the motion of electrons around the nuclei. This is the *van der Waals* force, and can cause chemically inert gases, like argon, to liquefy and solidify at low temperatures. But to understand how strong chemical forces bond atoms into molecules, we need to consider the way in which the more reactive atoms interact when their electron clouds overlap. The structure of the periodic table of Figure 1 is helpful. Atoms and molecules individually and in aggregates seek their lowest energy and most stable configurations, which correspond to the closed, or saturated, atomic shells of electrons. The eight electrons in the closed shells of atoms like neon and argon form a spherical ball of charge. All eight electrons occupy the same shell and enjoy an equally strong electrostatic force. The electron clouds are symmetrical and complete; this gives the noble gases no incentive to join up with other atoms to form molecules.

In moving along the periodic table one position from neon to sodium an extra electron is added, called a *valence electron*, which must go into a new outer atomic shell, making sodium highly reactive. If we move one space back from argon, we get to chlorine, which is one electron short of a filled shell and so also reactive. When sodium and chlorine atoms come together their outer orbitals overlap, enabling them to strike a mutually advantageous deal. The sodium can donate its outermost valence electron, which is easily ionized, to chlorine to complete its outer shell, and the whole structure becomes a stable molecule of common salt (sodium chloride or NaCl). Each atom benefits by achieving the stable closed-shell structure of a noble gas, and each is a charged ion, Na^+ and Cl^-; the ions are held together by an *ionic bond*. Salt forms a cubic ionic crystal with a lattice of alternating positive and negative charges. The *electropositive* elements that donate their surplus outer electrons, like sodium, are mainly metals. The *electronegative*, or electron acceptor, elements are found in substances containing oxygen, sulphur, chlorine, and fluorine.

Hydrogen is different. It has one electron in a shell made for two (helium), and so it can go either way, H^+, or H^-, in making an ionic compound. Hydrogen can accept an electron from say, lithium, to make lithium hydride, or share its electron to satisfy an electron-hungry element such as fluorine or oxygen. These weaker bonds are called *hydrogen bonds* and are common in many organic and biological molecules, for example between the base pairs that form the twisted double helix DNA molecule.

Two protons can join up by sharing two electrons to form a neutral hydrogen molecule, H_2. The two electrons form a strong *covalent bond*; the 'twoness' of the paired electrons comes from a quantum rule called the *Pauli exclusion principle* (see Chapter 5), which allows two and *only* two electrons to take part. Covalent bonds can therefore become *saturated*, for if a third hydrogen atom should approach the covalently bonded pair, it would be excluded and so cannot form a stable triatomic molecule.

The versatile element *carbon* has four electrons in its outer shell, and four vacancies. This structure enables carbon to form organic compounds with oxygen, hydrogen, and many others such as the biological molecules of life. We are 'carbon-based life forms', and our bodies contain polymer molecules. A polymer is a long-chain molecule, in which thousands of atoms or molecules line up, like beads on a necklace, with other atoms stuck to the sides to fill up the spare bonds. Many organic polymers have this chain-like form including common plastics such as polyethylene, a long chain of carbon atoms, with hydrogen atoms attached to the sides (Figure 7).

Some elements can exist in several different physical forms, called *allotropes*. Graphite, charcoal, and diamond are all allotropes of carbon. The extreme hardness and durability of diamond form arises from its 3D cubic arrangement of four strong covalent bonds, which contrasts strongly with the properties of its other allotropes such as graphite, which involves only three covalent bonds. Graphite has a layered structure with weak forces in between the tough layers making it slippery and able to write on paper; the 'lead' in a pencil is in fact graphite. There are other carbon allotropes too, and a large class of these are the *fullerenes*, which can form single-atom thickness carbon nanotubes, and hollow 60-atom-strong spheres of *Buckminster fullerene* ('buckyballs'). Carbon continues to surprise us, and in 2004, a new phase of carbon was extracted, *graphene*, which has remarkable mechanical and electrical properties. It is the thinnest possible layer of graphite, a sheet of carbon atoms, one atom thick, arranged in a 2D honeycomb lattice (Figure 8).

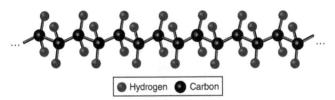

● Hydrogen ● Carbon

7. The chain structure of a polymer (plastic) molecule: polyethylene.

8. The atomic-scale 'chicken-wire' structure of graphene, a 2D sheet of carbon, one atom thick.

Crystals

The beautiful forms of gemstones or a snowflake (Figure 9) reveal the underlying symmetry of the arrangements of molecules in crystals. When large numbers of particles aggregate, they always try to minimize their potential energy, which draws them into 'close-packed' configurations. The minimum energy configuration of three atoms is a triangle. If more atoms are added in the same plane, they settle into a compact 2D hexagonal pattern, such as the graphite surface shown in Figure 4. This is easily demonstrated

9. The hexagonal beauty of a snowflake. It appears to be symmetrical, but it has less symmetry than the water vapour from which it condensed.

by filling a shallow tray with equal ball bearings. On shaking the tray, they will form a layer with a compact hexagonal pattern. A new layer of atoms can be added on top, fitting snugly into the hollows of the lower ones. With four atoms, we move into three dimensions, where the minimum energy configuration is a tetrahedron. As more atoms attach themselves to the seed crystal, they preferentially occupy the triangular faces and grow out to form larger hexagonal clusters. The efficient arrangements of the atoms and molecules in the *Hexagonal Close-Packed* (HCP) crystal structure is common to many crystals.

The hexagonal structure of the snowflake reflects the arrangement of its molecules. In a water molecule, two hydrogen atoms bond with two half-empty electron clouds that emerge at right angles from the oxygen atom. The hydrogen atoms prise apart the bonds slightly, to an angle of 105° so that the molecule is shaped like a shallow letter 'V' with the oxygen atom at the apex. This geometry gives water a unique set of properties. When water molecules form ice, their minimum energy configuration is a hexagonal ring, with a hole in the centre. The solid expands slightly on freezing which explains why water pipes burst and rocks split. The slightly lower density of ice also explains why icebergs float (90 per cent of their mass lies below the surface).

Intermediate states: glasses and liquid crystals

It would be easy to think that matter neatly divides up into just the solid, liquid, and gas phases. But nature is far more complex than that. Materials have a very wide range of different forms and I will highlight two examples: *glass*, a form that is neither solid nor liquid but a 'frozen liquid', and a *liquid crystal*, which has properties in between a liquid and a crystal.

When a liquid is cooled down to form a crystalline solid, the molecules must move from their close-spaced, disorganized configuration of the liquid phase, to the long-range ordered configuration and regularity of a crystal lattice. It takes time for the molecules to make these positional adjustments and, if the cooling is too rapid, they do not always have enough time to re-form into the long-range ordered configuration of the crystal. The molecules are in effect 'caught unawares' by their loss of mobility and get stuck before they can get to their crystalline positions. This is the vitreous, or glassy state of matter and one where there is a liquid-like disorganized molecular order, but the material has the rigidity of a solid. Solids, like glasses, that do not have definite geometric or crystalline structures are called *amorphous solids*, and in their disordered state there is not enough energy for the molecules to flow past each other. A glass is a 'frozen liquid', neither a regular solid nor a liquid, and is in a highly viscous metastable state, which means that over long periods of time glass will gradually tend towards crystallinity.

When a cooled monatomic liquid like argon freezes, the atoms form a close-packed regular lattice. Each atom can be imagined to be a hard sphere, which means that it has no preferred direction in space when settling into its minimum-energy configuration. But there are certain materials, called *liquid crystals*, which contain highly anisotropic rigid rod-like molecules that behave differently. Think of emptying a box of matches on a table. The matches spread out and can end up pointing haphazardly in any direction. If the matches are gathered together, they must all align in the same direction.

Similarly, the long molecules in a liquid crystal phase do not occupy regular positions on a lattice, but they can all be made to point in a single direction. A liquid crystal has the order and positional characteristics of a liquid, but it also has what is called *orientational order*. If the liquid crystal is heated, the rod-like molecules go back into the standard liquid phase, with the molecules pointing in all directions at random. A liquid crystal is therefore a state of matter that is intermediate between a liquid and an ordered crystal. These materials are the basis of the liquid crystal displays (LCDs) familiar in smartphones, computer screens, and TVs. Liquid crystals can be oriented by applying electric forces, moving easily from a disordered state to one where the molecules are lined up, a configuration that changes the way in which the material transmits light. In an LCD, this is achieved by applying voltages to the pixels of the screen.

The fourth state

Over 99 per cent of the normal matter in the universe is *plasma*, sometimes called the fourth state of matter. A plasma (Figure 5) is a gas that is so hot that it transforms to a new state. To understand how plasma differs from a gas, we have to think about the atoms that make up the gas. When we heat a gas to very high temperatures, the speed of the atoms becomes so large that their collisions knock off some of the orbiting atomic electrons, which are now free to move around on their own. The atoms that lose electrons are ionized, and are now positively charged *ions*. Plasma is a high-energy gas that is composed of two commingled populations of charged particles; the light negative electrons, and the heavy positive ions.

The name plasma comes from the Greek word meaning 'something moulded', and was coined by a pioneer of the subject, Irving Langmuir, in the 1920s. Two well-known naturally occurring terrestrial plasmas are lightning and the aurora or northern or southern lights. The aurora is produced when high-energy charged particles emitted by the Sun strike the Earth and ionize gases that glow high up in the atmosphere. Other familiar plasmas include candle flames, and neon and sodium lighting. Industrial and scientific applications of plasmas range from the manufacture of microcircuits to realizing the potential for unlimited clean energy from thermonuclear fusion power. Outside the atmosphere we are surrounded by a layer called the *magnetosphere* which is the plasma system formed by the interaction of the solar wind with the Earth's magnetic field. The Sun and all the stars are balls of hot plasma held together by gravity, and the solar wind is a stream of turbulent plasma that is blown away from the Sun's surface. Plasmas also form near exotic astrophysical objects like black holes. When matter falls towards a black hole, it settles first into a spinning disc surrounding the black hole. The friction in the gas heats it to very high temperatures, forming plasma so hot that it emits X-rays.

Plasmas can also be produced when atoms are ionized by high-energy particles, such as cosmic rays, or those produced by natural radioactivity. A small amount of natural ionization is present in most gases. Plasma can also form when electricity passes through gas. Air is normally a good insulator. However, if a very strong electric field is applied to it, as occurs during an electrical storm, it can 'break down' and become an electrical conductor. In a lightning strike, free charges are accelerated to high speeds before colliding with molecules of gas and knocking off charged fragments. This can generate a cascade of charged particles, amplifying and concentrating the ion–electron pairs to form a *plasma arc*, and carving out a conducting path through the gas by sheer brute force.

The most notable property of plasma is its electrical conduction, which results from the mobile charges it contains. The currents that flow in plasmas generate magnetic fields which exert forces that cause the plasma to 'pinch' and form narrow filaments. Such filamentary structures have been observed in the prominences and flares on the surface of the Sun. The electromagnetic properties of plasmas in terrestrial thermonuclear fusion reactors are exploited in various ways, both to heat them, and to confine them, often inside toroidal magnetic 'bottles'. (A torus is the shape of a ring-doughnut.) If the very hot plasma in such a device were to touch material walls, the impacts of the energetic plasma particles would erode them, by knocking atoms out.

One of the unique properties of plasma is that it exhibits *collective behaviour*. Examples of this are waves that can propagate through plasma without any particle collisions. In air, normal sound waves spread out when molecules collide, passing on their energy as compressions and rarefactions. In plasma there are equal numbers of positive and negative charges, making it electrically neutral to a very high level of precision. If this were not the case, the electric forces set up by even a small charge imbalance would move charges around until neutrality was obtained. Plasmas can support a range of different waves. Suppose for example that a sudden disturbance causes electrons to bunch up in some region of the plasma. The ions, being heavy and sluggish, cannot respond fast enough to restore the imbalance, and electric forces develop and push the electrons back. But the momentum of the electrons makes them overshoot their original positions, which they do until the electric force pulls them back again. Again, they overshoot. The cycle repeats, creating a disturbance that propagates through the plasma as *Langmuir waves*. The outermost part of the Sun's atmosphere, the solar corona, is very hot plasma, which supports these Langmuir waves. Plasmas host a wide range of other modes of oscillation, involving the collective motion of charged particles moving in electromagnetic forces.

In this chapter, we have seen the forces at work when atoms get together. The central idea is that aggregations of atoms will tend to drop into the lowest energy state that is available to them. There is a competition between interatomic electric forces, which tend to attract atoms and molecules together into clumps, and the dispersive effect of the thermal motion. In gases, the thermal motion wins and in solids the victors are the interatomic forces. A range of intermediate states can also exist, such as the 'frozen liquids' of glasses, and the orientational properties of molecules in liquid crystals. Most of the normal matter in the universe is in the fourth state of matter, high-energy plasma where particle energies are so high that electrons are knocked off the atoms and the matter exists as an intimately mixed gas of ions and electrons.

In Chapter 4, we move away from the Newtonian world and see how our understanding of matter was completely transformed by the great discoveries of electromagnetism and relativity.

Chapter 4
Energy, mass, and light

At the beginning of the 20th century, physics was turned on its head by two great revolutions: relativity and quantum mechanics. These changed forever our understanding of matter. In this chapter I will outline Einstein's *special theory of relativity* of 1905, which describes what happens when objects move at speeds close to the speed of light. The theory transformed our understanding of the nature of space and time, and matter through the equivalence of mass and energy. In 1916 Einstein extended the theory to include gravity in the general theory of relativity, which revealed that matter affects space by curving space around it.

To put things in perspective, we should first touch on the classical Newtonian picture. Newton had, by 1700, established laws of motion and the theory of gravitation. Provided speeds are not too high, and masses not too large, Newton's laws provide a very good framework for understanding the world, enabling us, for example, to put a man on the Moon. Newton's universe rested on two assumptions. First is the idea of an absolute *time*; his laws seem to contain the notion that there is a cosmic clock ticking away, which everyone in the universe would agree with, wherever they are. Second is the concept of an absolute and immutable *space*.

Newton was aware that, apart from gravity, there are other forces in nature, such as the electric force. Electrical charges attract or repel each other at large distances, a long-range property shared also by magnets. When you hold a pair of magnets in your hands, you can feel the repulsion of like poles and the attraction of opposite poles. It is not hard to imagine that the magnets are immersed in some sort of invisible 'force field'. Electricity and magnetism are deeply connected, a fact that was discovered by Danish physicist Hans Christian Ørsted in 1820 when he observed that a magnetic compass needle was deflected by a nearby wire carrying electric current. André-Marie Ampère went on to determine the force law between current-carrying wires.

Our understanding of the connection between electricity and magnetism took a giant leap forward in the 1830s with Michael Faraday's experiments on coils, batteries, and circuits. Faraday discovered that a changing magnetic field produces electric forces, an effect known as *induction*, which governs all practical electric generators. Faraday's experiments led him to a brilliant insight. What Faraday 'saw' in his mind's eye was the *electromagnetic field*, an invisible tension or stress that spreads out through empty space. The field makes its presence felt by producing forces that act on nearby susceptible bodies. Faraday imagined that charged or magnetic bodies produce a bundle of *lines of force*, sprouting from their surfaces (Figure 10). The lines transmit their forces to the bodies as if connected by invisible cables, pushing or pulling on them. The lines can be rendered visible when iron filings are scattered on a card placed over a bar magnet. A speck of iron is itself like a little magnet, lining up with the magnetic field, just as a compass needle aligns to that of the Earth.

Faraday's brilliant experiments showed that the lines of force spread out through space between the bodies on curved paths. This idea clashed with the way that Newton imagined how the gravitational force was transmitted instantaneously between two separated mass points directly along the line joining them. While Newton's theories had been immensely successful in explaining the motion of the planets, the concept of instantaneous 'action at

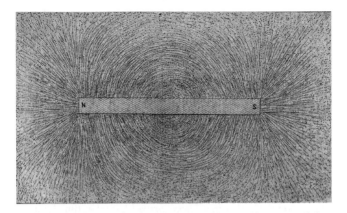

10. Michael Faraday's sketch of the magnetic lines of force, revealed by iron filings scattered over a bar magnet.

a distance' seemed out of character with most of the processes familiar to us from everyday life.

In a physical field every point in space can be labelled with a number representing the *field strength*, which varies from point to point. On weather maps, for example, temperatures or pressures are represented by a grid of ordinary numbers. Maps like these represent so-called *scalar fields*, where the field quantity is represented by a number associated with every point in space. There are also more complicated *vector fields*, such as a map of the wind velocity for which two numbers, speed and direction, are needed at each point. Weather maps are shown with arrows indicating the wind speed (the length of the arrow), and the compass bearing (its direction).

In 1864 Faraday's intuition about fields was put on a mathematical basis by James Clerk Maxwell in his famous set of equations that described electric and magnetic fields, unifying them into a single entity: the *electromagnetic field*. Maxwell used vector fields to describe how the magnitudes and directions of electric and magnetic forces vary in space and time. He also realized that fields spread out into empty space, disconnected from any matter. His equations showed that fields spread out at the velocity of light, and he guessed that light is an *electromagnetic wave*.

What is an electromagnetic wave? Imagine an electron, with its attached field lines sprouting out, being shaken rapidly to and fro. What happens to the electric field? Close to the electron, the field lines adjust rapidly to its changing positions. But it takes longer for the information about the varying position of the electron to reach points further away. The information spreads out in space, something like the way that buckets of water are passed from one person to another in a '*bucket brigade*'; it takes time for a bucket to be passed down a line of people. Maxwell's equations predict that when an electron is shaken, the oscillating electric field generates a complementary oscillating magnetic field which in turn generates an oscillating electric field, and so on. The two interlinked fields move out through empty space as a single undulating entity, transporting energy with them. The equations contain two easily measured physical constants, and when Maxwell put these into his theory he discovered that it predicted that the waves travel at a fixed speed in a vacuum, the speed of light, c. In fact, light *is* an electromagnetic wave. Maxwell's theory was the most important scientific discovery of the 19th century; the great triumph was that in one stroke Maxwell had unified three branches of physics: electricity, magnetism, and optics.

There was more. The Maxwell equations also predicted that the field should vibrate with a much wider range of wavelengths than just those of visible light. Our eyes have evolved to sense the narrow band of wavelengths in sunlight. However, there are longer wavelengths beyond the red end of the spectrum and shorter ones beyond the blue that we cannot see. There is a vast *electromagnetic spectrum* from the short wavelength of gamma rays (equal to the diameter of a proton) to radio waves many thousands of kilometres long. Maxwell's prediction of the electrical nature of light was confirmed soon afterwards by Heinrich Hertz's brilliant experiments on the generation and detection of radio waves.

The speed limit

While Maxwell's theory was tremendously successful, it clashed in a subtle way with Newton's ideas. Here is a thought experiment. First the Newtonian view. Suppose you are travelling on a bus at a steady 70 km/h and you throw a ball forwards at 10 km/h. From your point of view in the frame of the bus, the ball travels at 10 km/h. But if a roadside observer measures the ball's velocity, they would find it to be $70 + 10 = 80$ km/h. In Newton's world, the concept of a velocity has a meaning only when it is measured with respect to another velocity. A velocity adds when you are moving towards a body, and subtracts when you are moving away from it; Newtonian velocities are *relative*.

Next consider what happens when the bus driver switches on the headlights. If you measure the speed of the light on the bus, you'd find it to be c. What would our roadside observer find? They would *not* measure it to be $70 + c$, but *still* measure the speed of light to be c, the same as you measure it to be. This raises a question: is there a *velocity* that light is moving relative *to*?

One way out of the problem of the addition of velocities was the possible existence of the *aether*, a hypothetical all-pervasive light-bearing medium through which light propagates, but a substance that does not interact with matter. All common waves need a medium for propagation: for example, sound waves are transmitted through the air, and ripples move across the surface of water. When we gaze up at the stars, their light has travelled great distances through the vacuum of space. If the aether exists, then the speed of light should depend on the motion of the Earth through it. The fact that sound waves propagate faster downwind than upwind prompted a famous experiment by Albert Michelson and Edward Morley in which they attempted to measure the speed of light along and across the Earth's path around the Sun. The experiments always yielded the same answer: c is a constant. Experiments to detect the dependence of the velocity of light on the motion of the observer all failed.

In trying to make sense out of the baffling non-existence of the aether, Dutch physicist Hendrik Lorentz examined the Maxwell equations, and in particular the way they change when expressed in different inertial frames. An *inertial frame* is a grid of coordinates that is moving at some constant speed in a straight line with respect to some other inertial frame. Lorentz found that the equations take on different forms when formulated in different frames, which would imply that the speed of light should change in going from one to another, a prediction that clearly contradicted the negative Michelson–Morley experiment. Lorentz defined a mathematical transformation (the *Lorentz transformation*), which enabled the Maxwell equations to have the same form in different inertial frames.

This was the problem that Einstein, a 26-year-old clerk in a patent office in Bern, attacked with iron logic in 1905. Einstein interpreted the Lorentz transformation as expressing a profound physical relationship between space and time for observers in different inertial frames. His *special theory of relativity* was underpinned by two ideas: namely that the laws of physics should be the same, and that the speed of light in a vacuum is always the same for observers in different inertial frames. Einstein was famous for his thought experiments, and one that he performed when constructing his special theory was to ask what the world would look like if he could ride on a light beam; he would not be able to see his own image in a mirror, because light would never leave his face.

If the speed of light is the same for all inertial observers, then our notions of space and time have to be adapted and merged to allow for it. Einstein replaced Newton's concepts of absolute space and absolute time with a single merged entity: the fabric of four-dimensional *spacetime* (three space dimensions plus time) in which space and time become elastic and change for different observers. For example, if you have a ruler travelling at great speed past you, you would see the ruler shortened along its direction of motion; this is called the *Lorentz–Fitzgerald contraction*. Also, if there were a clock moving uniformly at high speed past you, you would see it running slow; this is called *time dilation*. Time dilation has been very well tested in laboratory experiments and is a consequence of the fact that all bodies are moving through spacetime at lightspeed. The special theory of relativity in effect completed Maxwell's theory. Einstein said that Maxwell was the only one of his predecessors who was on a level with Newton.

But the most far-reaching prediction of the special theory came from considering how Newton's laws of motion had to be modified for bodies moving at high speeds. The speed of light is nature's ultimate speed limit, and no material body can move faster than the speed of light. Newton's second law tells us how a body's velocity changes in response to a force, which is the rate of change of momentum. So, if a constant force were applied to a mass, there would be no limit to the velocity that it could attain, even exceeding the speed of light, which would violate relativity. Newton's formula for the momentum of a particle therefore had to be modified, with a relativistic transformation that led to the most famous equation in physics, the law of the equivalence of mass and energy:

$$E = mc^2.$$

The equation tells us that the mass m and energy E are really the same thing; they are just measured in different units (the c^2 factor simply converts mass units to energy units). If an object is travelling very fast, it carries a lot of energy with it, but there is a limit to how much it can have. Even when the object is stationary it has a 'latent' amount of energy called the *rest-mass energy*, and this is the real meaning of the '*m*' in the equation. This is different from Newtonian mechanics where the kinetic energy of a stationary particle is zero, not a finite quantity. The rest-mass energy is enormous for matter because the conversion factor c^2 is so large. Even a small mass contains a huge amount of energy. The energy equivalent of one gram of matter (the weight of a business card) is 25 million kilowatt-hours, enough to heat and light a large city for a day. Why don't we notice the energy content of ordinary matter in everyday life? If none of it is given off externally, none can be observed. Einstein asked us to imagine a very wealthy but parsimonious man. If he never spends a single coin how would we know that he is fabulously rich?

Matter as mass-energy

Einstein's famous result was announced in one of his four *annus mirabilus* papers of 1905. It has the title '*Does the inertia of a body depend upon its energy-content ?*' and in it he put the mass on the left-hand side of the equation:

$$m = E/c^2.$$

While it is mathematically identical to the better-known form, put this way the formula answers the question in the paper's title by defining mass in energy units. For example, the rest masses of an electron and a proton are respectively 0.511 MeV/c^2 and 938.25 MeV/c^2. (The electron volt unit of energy, eV, is defined on p. 55.)

It is easy to add some energy to matter to increase its mass, for example by heating it. If you boil a kettle of cold water the relativistic mass increase is tiny (about 5×10^{-13} kilograms). Even in energy-liberating chemical reactions, as in burning fuel, again only a negligible fraction (1 part in a billion) of the mass is released as energy. Lavoisier's law of mass conservation is therefore a very good approximation for chemical processes. But in nuclear reactions much more energy is released and Einstein's equation is seen as the most awesome one in science because it is the basis of nuclear power.

The nuclear reactions to which we owe our existence are the fusion reactions that take place deep inside the Sun. There, the temperature is high enough to enable four protons to fuse together to produce an *alpha particle* (the nucleus of a helium atom), which has a rest mass of 3727.3 MeV/c^2. In forming the alpha particle, the protons lose a significant amount (0.7 per cent) of their mass, and the deficit, 26.7 MeV, goes into heat energy. An alpha particle is nearly 1 per cent lighter than the sum of its parts.

Even higher rest-mass energy fractions are released in the most spectacular events in the universe, for example when black holes merge. Several mergers of black hole pairs have now been observed via their gravitational radiation signals, which were first observed in 2015. These mergers can involve black holes with masses of tens of solar masses (a solar mass is the mass of the Sun). The final stages of the mergers occur in a few tenths of a second and create such violent ruptures in the fabric of spacetime that the equivalent of a few solar masses is converted into gravitational wave energy. There is an even more extreme process involving *antimatter*, which we will look at in Chapter 6. When matter encounters antimatter, it annihilates completely. All of the rest-mass energy is converted to radiation with 100 per cent efficiency; in science fiction this process has been envisaged as the propulsion system for the *Starship Enterprise*.

Curved spacetime

Newton's great discovery of the laws governing the motions of the planets hid a profoundly subtle and remarkable fact about the nature of mass. Newton calculated the orbit of a planet by combining his second law of motion (force = mass × acceleration) with the law of gravity (the force between two masses is proportional to the mass of one × the mass of the other). But there are two different types of mass involved here. The mass that appears in the law of motion is connected with a body's resistance to a change in its motion, its inertia. This is its inertial mass, which has nothing to do with weight. The inertial mass of a body is the same whether it is on the surface of the Earth or is in deep space far away from any planet. But the mass that appears in the law of gravity is connected with gravity and weight. An object with more mass is attracted more strongly to other masses by gravity. In this case the mass is called *gravitational mass*, and can be thought about as something like the 'charge' of gravity.

It was implicit in Newton's formulation of planetary motion that inertial mass and gravitational mass are one and the same thing, an assumption that is by no means obvious. Einstein elevated the equality of the two types of mass to a principle: *the principle of equivalence*, which he took to be the foundation of his general theory of relativity of 1915. It is one of the great achievements of science. In this, he recognized the significance of Galileo Galilei's famous experiment of dropping a light ball and a heavy ball, which are observed to fall with the same acceleration in the gravity of the Earth. But imagine repeating the same experiment, this time inside a rocket ship out in space, far away from any gravitating matter. If the ship fires its engine and starts accelerating, the dropped masses would appear to accelerate towards the floor. We can imagine that the objects are at first just floating in space and, when the rockets are fired, the floor approaches *them* at an ever-increasing speed. To an on-board observer, this motion mimics the acceleration of gravity. The observer therefore cannot tell if the ship is being accelerated, or if a large gravitational mass has suddenly appeared behind the vessel.

Matter

It took Einstein ten years to incorporate the effects of gravity and acceleration in the more complex and mathematically demanding general theory. To do this he had to abandon the notion that spacetime can be described by regular Euclidean geometry and, in the presence of gravitating masses, must instead be based on the geometry of curved surfaces. What does it mean to say that spacetime is curved? In the plane geometry of flat Euclidian space, the angles of a triangle add up to 180°. But if you draw a big triangle on the surface of the Earth, you find that the angles don't add up to 180°, because of the curvature of the Earth. If you imagine drawing a triangle in space, near an object with a gravitational field, the geometry of space itself is curved and here you would again find that the angles of the triangle don't add up to 180°.

The principle of equivalence led to a profound view of gravity, as the manifestation of the geometry of space and time. The fact that different mass bodies falling in a gravitational field all accelerate by the same amount and so follow the same paths reveals a deep truth: the *paths* that the bodies follow are themselves inherent features of spacetime. Matter curves space around itself. Imagine a rubber sheet with an object with large mass-energy like the Earth sitting in the middle making a 'dent' in it (Figure 11), which represents the curvature of space around a massive body. The Moon feels the curvature of space and moves in its orbit around the Earth, as if it were a marble rolling around inside a bowl. The concept of Newton's force of gravity was replaced by the curvature of spacetime.

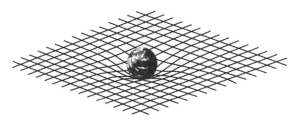

11. **The curvature of space round the massive body of the Earth.**

The curvature of space around a mass not only affects the motion of masses, but it also deflects light. In 1919 Arthur Eddington famously put Einstein's theory to a key test during a total eclipse of the Sun. The apparent positions of stars observed near the edge of the Sun were seen to be displaced from their normal positions on the sky by precisely the amount that was predicted by Einstein's theory.

General relativity is the correct theory to use in describing what happens near matter that has suffered extreme gravitational collapse. The ultimate light-bending limit occurs in the immensely strong gravitational field of a *black hole*, the most compact form of matter we know of. If all the matter in the Sun, which has a diameter of 1.4 million kilometres, were compressed into an object just 3 kilometres across, the gravitational field at the surface would be so enormous that spacetime would wrap itself around the object and a black hole would form. Light would be bent so strongly at the surface that it would be completely trapped, as would any material object falling into it. We can rest assured that other processes will prevent the Sun from collapsing into a black hole, but this example serves to show how extreme the physical conditions would be, as well as the inherent weakness of gravity. Spacetime is so massively distorted around a black hole that time itself appears to be frozen at the surface. The effect of gravity on the time indicated by clocks is even important for satellites orbiting in the weaker gravity of the Earth. The *global positioning system*, or GPS, on which we depend for our navigation, would not work if general relativity were not considered.

To summarize; in the 19th century, electromagnetic forces came to be understood as electric and magnetic fields filling empty space which couple together and oscillate as waves. In fact, light *is* an electromagnetic wave. Einstein built the special theory of relativity on two postulates: that the speed of light is the same for all observers, and the laws of physics are the same for observers moving at uniform speeds. From these, he deduced that space and time merge into a single fabric: spacetime, in which space and time are distorted when seen by different observers. Newton's laws of mechanics also had to be modified, which revealed that a central property of matter, its mass, is equivalent to energy. When Einstein included accelerated motion and gravity as the basis of his general theory of relativity, acceleration and gravity were found to be equivalent. The geometry of spacetime is determined by the distribution of mass-energy in space. The physicist John Wheeler put it succinctly: *matter curves space, and space tells matter how to move.*

Einstein's general theory of relativity revolutionized physics. In Chapter 5, I will turn to the world of very small particles of matter and the second great revolution in 20th-century physics: quantum mechanics.

Chapter 5
The quantum world
of the atom

By the end of the 19th century, the physical laws describing the behaviour of macroscopic chunks of matter had largely been established. These included Newton's laws of motion and gravity, which could predict phenomena ranging from the collisions of billiard balls, to the motions of the planets. There were also the laws of thermodynamics, Boltzmann's statistical mechanics, and the unification of electricity, magnetism, and optics in Maxwell's laws of electromagnetism. But the structure and the inner world of atoms and molecules remained a mystery. There were unresolved puzzles, such as the origin of sharp atomic spectral lines, seen in the light from the stars. The wavelengths of these lines coincided tellingly with those emitted by chemical elements heated in laboratory flames. This chapter charts the story of how, in parallel with profound discoveries of the properties of atoms, the revolutionary theory of quantum mechanics was born.

Several important clues about the structure of atoms had already been uncovered, particularly during the final decade of the 19th century. These included the highly penetrating X-rays (by Wilhelm Röntgen) and radioactivity (by Henri Becquerel). Radioactive atoms, such as uranium, fire off parts of themselves at great speeds, disintegrating spontaneously and emitting different types of ionizing radiation or particles. One kind of radioactivity, *gamma rays*, is high-energy electromagnetic radiation. Another kind consists of positively charged alpha particles, which are very stable entities, and were shown by Ernest Rutherford to be helium nuclei ejected at speeds of 10,000 km/s (about 30,000 times the speed of sound), and the third (*beta particles*) are high-speed electrons.

The heart of matter

To probe inside the atom, Rutherford hit on the idea of using the heavyweight alpha particles as atomic 'bullets'. In 1912, with Ernest Marsden and Hans Geiger, he fired alpha particles at a thin gold foil (Figure 12). Most of the fast particles ploughed straight through the millions of gold atoms they encountered, brutally shoving aside the lightweight electrons. But occasionally the alpha particles were scattered through small angles, fewer were deflected by larger angles, and a small fraction were even scattered *backwards*. In Rutherford's own words: 'it was as if you had fired a 15-inch shell at a piece of tissue paper and it came back and hit you'. The experiment showed that in the centre of the atom is a tiny, positively charged, and heavy nucleus. Occasionally an alpha particle might strike a gold nucleus head-on, feel its strong electrostatic repulsion, and rebound. The single proton nucleus of a hydrogen atom has a diameter of 10^{-15} m, about 100,000 times smaller than the atom itself. Atoms are mostly empty. If a

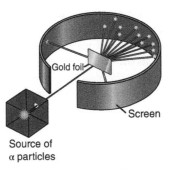

Gold foil

Screen

Source of
α particles

12. Rutherford's experiment to scatter alpha particles from gold atoms led to the discovery of the atomic nucleus.

hydrogen atom were enlarged four trillion times to the size of London's Wembley stadium, its nucleus would be the size of a pea.

The nucleus is small, massive, and positively charged, but of what was it made? A hundred years earlier, Dalton's chemical experiments had revealed the simple number ratios of the atomic weights of the elements, in units of the weight of the hydrogen atom. It was guessed that nuclei were made from the building blocks of protons, the nuclei of hydrogen atoms. In 1919 Rutherford confirmed this by knocking protons out of nitrogen, again using alpha particles.

The discovery that the nucleus contains protons raised a deeper question. In moving along the periodic table, the atomic weight of an element increases much faster than its atomic number. For hydrogen the atomic number and weight are both 1. In helium (atomic number 2), the weight is four times larger. By the time we get to the 92nd element, uranium, the atom weighs about 238 times as much as hydrogen. Something was needed to make up the deficit between the charge and the mass in the nucleus, and Rutherford guessed that, in addition to protons, other heavy particles must be present. He called these *neutrons* since they carry no charge. James Chadwick discovered the neutron in 1932 when he fired alpha particles at beryllium atoms and knocked neutrons out of them. A neutron weighs a tenth of a percent more than a proton; but apart from its electrical neutrality, the two are similar. The protons and neutrons in a nucleus are called *nucleons* (Figure 13). Elements with the same number of protons in their nuclei all have the same chemical properties, but they can have different numbers of neutrons, which makes them *isotopes* of that element. For example, helium-4 has two protons and two neutrons whereas helium-3 has two protons and only one neutron.

How is the nucleus held together? With up to around 100 protons crammed into its tiny volume, the mutual electrostatic repulsion in the nucleus is enormous and a different force, the *strong nuclear*

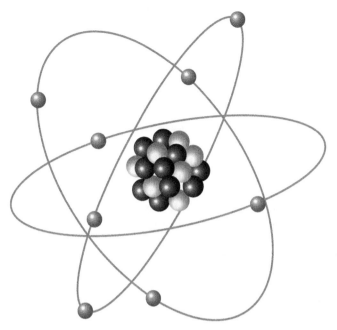

13. Representation of an atom showing the neutrons and protons clustered together in the atomic nucleus, surrounded by orbiting electrons. (Caution: the image of the nucleus has been magnified by over 100,000 times relative to the electron orbits; if it were to scale, it would be invisible.)

force, binds the whole cluster of particles together. The strong force is much more powerful than the electrical force, and operates over a much shorter range of around 10^{-15} metres. We can appreciate how strong this force is by comparing it with how tightly an electron is bound to an atom. The energy unit that is conventionally used on the microscopic scale of matter is the *electron volt*, or eV. The work needed to move an electron through an electric potential, say between the terminals of a 1.5-volt 'AA' battery, is 1.5 eV of energy. (There are 6.25×10^{18} eV in one Joule.) To remove an electron from an atom, it takes about 15 eV of energy, which could be supplied by 10 AA-size batteries. But to remove a proton from a nucleus it would take about 10 MeV of energy, which is equivalent to *seven million* batteries. If you could reach into an atomic nucleus with a strong pair of tweezers and pluck out a neutron, the force needed would be about the same as lifting a heavy suitcase.

Quantizing the atom

The discovery of the atomic nucleus led to a picture of the atom as a miniature solar system, not unlike Figure 13, in which the nucleus sits in the centre like the Sun, with the electrons orbiting around it like planets. However, it was quickly realized that if the electrons behaved classically, atoms would rapidly radiate away all their energy and collapse under the attractive force between positive and negative charges. Where does the stability of atoms come from?

The key concept that was needed to understand atomic stability had already been proposed in 1900, but in a different area of physics—the study of the radiation emitted by hot bodies. In studying thermal radiation, German physicist Max Planck had a radical idea: radiation energy must be *quantized*. Planck proposed that the energy of the electromagnetic field is bundled up in discrete packets, or *quanta*, rather than forming waves with a continuous range of energies right down to zero, as predicted by Maxwell's theory. The energy of Planck's quantum is proportional to the frequency of the radiation, so that a quantum of high-frequency blue light carries, for example, roughly twice as much energy as one of red light. The constant of proportionality, h, is a fundamental constant of nature, called *Planck's constant*. It has the same units as the momentum of a spinning body, or angular momentum. Planck's constant is a very small number and defines the smallest steps taken by nature, the *quantum of action*.

Soon after Planck's revolutionary proposal, Einstein extended the quantization idea by showing that light travelling through space consists of particles, called *photons*, carrying discrete packets of energy. This idea explained a well-known phenomenon called the

photoelectric effect in which metals such as zinc can be made to
eject electrons by shining high-frequency ultraviolet light on them.
To prise an electron out of a metal, a minimum amount of energy
has to be delivered to it, and for this the high frequency of the
photon is crucial. If the frequency of the light is below a critical
value no electrons are emitted however intense the light source.

The Danish physicist Niels Bohr applied the idea of the
quantization of energy to the electronic structure of atoms.
He theorized that an electron moving in a circular orbit around
the nucleus could only occupy discrete orbits with whole number
values (1, 2, 3...) of the angular momentum, in units of $h/2\pi$
The whole number values are called *principal quantum numbers*.
The energies of the different orbits form a discrete *ladder* of
allowed states with unequally spaced rungs (Figure 14). The

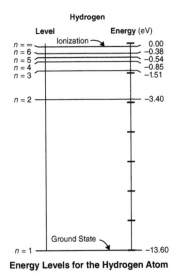

Energy Levels for the Hydrogen Atom

14. The ladder of quantum energy levels of the hydrogen atom shown as
a series of horizontal lines. The energies are negative numbers because
they represent the energy needed to remove an electron from a given
level to infinity. The principal quantum numbers are labelled by n.

quantum rule is that the electron in an atom is allowed to sit *only* on the rungs of the ladder, but never in between them. It can jump up to a higher rung (an *excited state*) by absorbing a quantum of energy equal to the difference in energy between the rungs, or fall down to a lower energy rung by emitting a photon with a specific energy and therefore frequency. The lowest rung on the ladder is called the *ground state*, and, because there are no states below it, electrons are prevented from crashing into the nucleus; it is this that makes atoms stable. In addition to explaining the stability of atoms, Bohr's model also resolved the long-standing question of the origin of the sharp atomic lines in atomic spectra. When he used the theory to predict the observed wavelengths of the hydrogen line spectrum, his model of the atom was immediately hailed as a major triumph.

Particles and waves

Nature imposes strict quantum rules on matter. An electron cannot have any energy in the atom, but is constrained to a ladder of discrete quantum energies. How is it possible to make sense out of this strange behaviour? The key is the realization that microscopic particles possess wavelike characteristics. Waves of all types have basic features such as wavelength (the distance between successive wave crests) and amplitude (the height of the wave). They can propagate freely through space as travelling waves, or as stationary or standing waves, for example the vibrations of a guitar string. Electrons in atoms behave like standing waves.

The quintessential property of waves is *interference*. If you throw two stones simultaneously into a still pond, the circular ripples from each spread out and cross. Where the waves meet, each adds to the other (called *linear superposition*). Where two wave crests meet, they add to make a large double height hump (*constructive interference*); when two troughs meet, they make a trough of double depth. If a crest and a trough meet, they cancel each other out so that the water surface is level (*destructive interference*). Light shows these interference patterns, and in 1801 Thomas Young performed his famous double slit experiment (Figure 15), which established decisively the wave nature of light. When light is shone on a pair of closely spaced parallel slits cut in a screen, the waves spread out on the far side to produce a pattern of bright and dark interference bands of constructive and destructive interference.

After Einstein's explanation of the photoelectric effect in terms of photons it had become clear that light has both a particlelike and wavelike character. Wave–particle duality was (and still is) a hard concept for us to digest; but the evidence for the existence of photons and Young's wavelike interference patterns is irrefutable. In 1924 an idea occurred to a young French aristocrat, Prince Louis de Broglie. If light has a dual nature, then why shouldn't

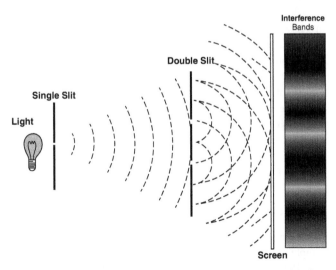

15. Thomas Young's double slit experiment showing bright and dark interference bands or fringes which are produced by the constructive and destructive interference of waves emerging from the slits.

the smallest particles of matter also have a wavelike character? De Broglie defined the wavelength of a quantum particle, the quantum *de Broglie wavelength* λ_{dB}, in terms of its momentum mv (mass × velocity):

$$\lambda_{dB} = \frac{h}{mv}.$$

The size scale on which wavelike quantum effects are important in matter is fixed by the magnitude of Planck's constant, in the numerator. The particle's momentum appears in the denominator, which predicts that the more massive a piece of matter is and/or the higher its speed, the smaller is the quantum wavelength. This means that macroscopic objects, like billiard balls, have de Broglie wavelengths that are unnoticeably small. But on the scale of atoms the wave nature of matter is central.

De Broglie conceived of waves that correspond to freely moving particles. But, when viewed as standing waves, they give insight into how Bohr's electron orbits are quantized. If one imagines bending an electron wave into a circle around a nucleus, there are only certain ways that this can be done so that its ends join up smoothly. The mathematical condition is that the circumference of the orbit must contain an exact number of wavelengths. If either the wavelength or the circumference deviates even slightly from this condition, the electron wave would not join up smoothly and would rapidly get out of step with itself and cancel out. This is the reason why electrons cannot sit in the gaps between the rungs on the quantum energy ladder.

The uncertainty principle

In classical mechanics, Newton's laws of motion describe completely the motion of a macroscopic particle moving in a force field. Both the position and the momentum of a point-mass are mathematically well-defined and independent quantities and its trajectory is sharply defined everywhere in spacetime. But this is not true at the quantum level where particles travel from one location to another along every possible path through spacetime. It makes no sense to think about pinpointing an extended quantum object to a region of space smaller than its own wavelength. Furthermore, the wavelength of a quantum object varies inversely with its momentum according to de Broglie's formula, and the position and momentum variables are not independent. A large matter wavelength implies a small momentum and vice versa. Position and momentum variables form special pairs, called *complementary variables*.

In 1926 German physicist Werner Heisenberg constructed the first complete theory of quantum mechanics (called matrix mechanics, involving mathematical objects called matrices). In this he focused on what it means to make a measurement of a quantum system and concluded that the knowledge we can obtain about it is fundamentally limited by nature, with a precision that is not related to any imperfections or limitations in the measuring apparatus. A quantum entity doesn't *itself* know where it is and how fast or in what direction it is going—the quantum world is inherently indeterminate. Heisenberg hit upon the key idea, the *uncertainty principle*, which asserts that we can know either where a quantum object is or where it is going, but we can't know both at the same time. The principle underpins the whole of quantum mechanics and tells us that the quantum entities in it are neither pure particles nor pure waves.

The uncertainty principle links the spread or uncertainty (indicated by the Δ symbol) in a particle's position (Δx) with that of its momentum (Δp), so that the product of the two is greater than or equal to Planck's constant/4π:

$$\Delta x \, \Delta p \geq h \, / \, 4\pi.$$

How does this work? Suppose we want to measure the position and the speed of an electron. We can shine a light on the electron and find out where it is and how it moves by the light that it scatters back to us. However, the electron has a very small mass and, to avoid disturbing it too much, we need to shine only a dim light on it. However, the electron must scatter at least one photon for it to be seen at all; when it does, there is an exchange of energy and momentum. The photon therefore inevitably imparts momentum to the electron, an interaction that makes it fundamentally impossible to observe a particle without disturbance. What are the options? We can choose to disturb the electron as little as possible and so make a more precise measurement of the electron's *momentum* by using a low-energy long-wavelength photon, illustrated on the left of Figure 16. But when we do that, since we cannot pinpoint an object in space with a precision that is *smaller* than the wavelength of the light used (in this case large), our measurement of the *position* becomes uncertain. Alternatively, we can choose to reduce the uncertainty in measuring the electron's *position* by using a high-energy

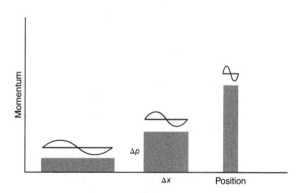

16. Illustration of Heisenberg's uncertainty principle, in which the quantum of action has been 'squeezed' in different ways. The uncertainty in measuring the momentum of a particle, Δp, is inversely proportional to the uncertainty of its position, Δx, so that the *area* in momentum-position space (indicated by the greyed-out rectangles) is proportional to Planck's constant. The long, medium, and short wavelengths of light used to measure a quantum particle are indicated above each rectangle, and described in the text.

short-wavelength photon (such as a gamma ray photon), as shown on the right of the figure. But, with a high-energy gamma ray photon, the 'kick' received by the electron is now so large that the *momentum* measurement becomes very uncertain. The uncertainty principle makes it impossible to refine both position and momentum measurements simultaneously, to arbitrarily high degrees of precision.

The uncertainty principle is very powerful. Among its many important consequences is the property that when a quantum particle is strongly localized in space it experiences large fluctuations in momentum and therefore kinetic energy, known as *zero-point energy*. Quantum entities constantly fluctuate in their lowest energy state. The uncertainty principle denies matter ever quite reaching the absolute zero of temperature, because, if it did, the particles would have precise locations in space, which is forbidden.

The concept of quantum localization energy also helps us understand the stability of atoms. Imagine building a hydrogen atom by bringing an electron and a proton together from infinity. The starting point is two widely separated particles, and the end point is the quantum-restricted motion of an electron moving in the electrostatic force field of its proton nucleus. As the electron approaches the proton, the kinetic energy it acquires from the electrostatic field makes it jitter about increasingly wildly, surrounding the proton as a vibrating fuzzy ball. At first, the ball is large, but as the electron radiates away its energy it shrinks down, and it becomes increasingly confined to the vicinity of the proton, occupying a restricted volume defined by size Δx. As Δx gets smaller, the spread in its momentum Δp must increase in line with the uncertainty principle. In turn, the kinetic energy of the electron increases, which opposes any further shrinkage. Eventually the competing effects of electrostatic attraction and the quantum localization energy balance each other as the electron relaxes into its minimum energy ground state. Nature's compromise is to produce stable atoms with quantum ladders of electronic energy levels.

Interpreting the quantum world

Shortly after Heisenberg published his matrix theory, Austrian physicist Erwin Schrödinger published an alternative approach, which was based on an equation describing the dynamics of wavelike quantum particles moving in a force field. The *Schrödinger equation* is similar to classical wave equations governing the way ripples move over a pond or sound waves propagate through the air. It is the fundamental equation of quantum theory and the analogue to Newtonian dynamics in the microworld. The Heisenberg and Schrödinger formulations, although outwardly different, are physically equivalent.

Solutions to Schrödinger's equation are called *wavefunctions*, and describe waves that are labelled with a symbol Ψ. This immediately raised the question of how the wavefunction should be interpreted. All waves have positive and negative values (for example sea waves can swell above or dip below mean sea level). So, if a quantum particle is represented by a wave, how should one interpret negative values? Max Born asserted that the *square* of the wavefunction Ψ^2 (which is never negative) represents the *probability* (and not the certainty) of finding the particle in a particular region of space.

A second essential feature of Schrödinger's equation is that it is *linear*. In a linear equation, the *sum* of two or more solutions is itself a solution. This means that the sum of wavefunctions is also a wavefunction representing a *mixed state* that permits a quantum system to be in *more than one state at a time*. Each quantum state can evolve independently, as if the others were not there. This feature gives rise to a bizarre and unique feature of the quantum world, *quantum superposition*. When a quantum system is in a *superposition of states*, it is not possible to specify its physical characteristics.

Quantum superposition can help us make sense out of the odd behaviour of how an atomic electron can jump between rungs on the quantum ladder, seemingly vanishing from one level (state A) and popping up on another (state B), without passing through any intermediate states. When the atom makes a quantum jump from state A to state B, the electron interacts briefly with a photon. During that time, the wavefunctions of the initial and final states are superposed, and the electron is said to be in a superposition of states. The superposition is a mixture of two states: the electron is in state A, *and* the electron is in state B. As the atomic transition unfolds, the wavefunction for state A gets weaker, while B's gets stronger until finally it alone remains. A rough analogy to this is the transient 'squeak' that is produced by a clumsily blown wind instrument, when it flips between two modes of oscillation.

There is a key experiment which gets to the heart of the superposition principle: Young's double-slit experiment, performed with particles instead of light. This experiment was a favourite of Richard Feynman's, about which he said: 'in reality it contains the *only* mystery ... of all quantum mechanics'. Let's look again at the experiment of Figure 15, using a beam of electrons fired at a metal screen containing two parallel slits. On the far side is a glass screen. If electrons pass through the slits, they will strike the glass screen, and show up as tiny flashes of light. If one of the slits is covered up, the pattern of flashes on the glass screen is found to be uniform, with the electrons passing through the open slit and hitting target points on the far screen as if they were tiny bullets. If the blocked slit is now uncovered, the electrons can pass through *both* slits and the pattern of flashes on the glass screen changes radically. The uniform pattern now switches to a wavelike pattern, with bands of constructive and destructive interference. Significantly, electrons are now being *denied* access to certain regions of the screen that they could readily hit when only one slit was open. This dramatic result completely contradicts our mental picture that electrons behave like tiny bullets; why, if electrons are like bullets, should opening a second slit influence the electrons passing through the first?

This experiment can be taken one stage further. The firing rate of the electron gun is now turned down so that there is only one electron flying through the apparatus at any time. With both slits open, the positions of the electron hits on the glass screen are recorded and, over time, gradually build up a wavelike interference pattern, identical to the one of the earlier experiment, when both slits were open. But how can this be if there is only one electron in the apparatus? Does an electron spread out and somehow pass through both slits at the same time in order to interfere with itself? If we try to identify through which slit an electron passes, say by scattering a photon from it, the act of measurement disturbs the state of the electron and the pattern on the glass screen immediately flips back to the uniform one.

Let's summarize this weird behaviour. If nobody is observing the electron, it can apparently go through both slits at once. In that case the electron wavefunction is in a superposition of states: namely, the electron goes through the first slit, *and* the same electron goes through the second slit. But if the electron is observed, its wavefunction is said to *collapse* into a single state corresponding to it going through only one slit. In effect, looking and not looking at the electron creates two different experiments, each of which produces different results.

The concept of the *collapse of the wavefunction* is central to the '*Copenhagen interpretation*' of quantum mechanics, a set of ideas put forward by Niels Bohr in 1927. The name comes from the location of Bohr's institute. Bohr recognized that our knowledge of the quantum world comes only from the measurements we make at the macroscopic level, using typical laboratory apparatus. Because a quantum system is disturbed when measurements are made on it, it is meaningless to ask what a quantum particle is doing when no one is looking at it. The only thing one can do is to calculate the *probabilities* that a quantum system occupies certain states. The instant that an observer makes a measurement, the system is forced to collapse into a unique state, whereupon the system 'decides' what state it is in. Einstein was never happy with the probabilistic basis of the Copenhagen interpretation and, in a famous series of exchanges with Bohr, famously announced that 'God does not play dice', to which Bohr replied: 'Einstein, stop telling God what to do.' It turns out that however we choose to interpret the wavefunction, the Schrödinger equation has passed exacting experimental tests.

But even Schrödinger was not happy with either his theory or the Copenhagen interpretation. To illustrate the absurdities that arise, he concocted a hypothetical thought experiment called *Schrödinger's cat*. He imagined placing a cat in a sealed box, along with a radioactive atom and a device to release a deadly poison gas the instant that the atom disintegrates, an action that would kill the cat. After a certain time, the chance that the atom has disintegrated is 50 per cent, and the atom is in a superposition of two states: not disintegrated *and* disintegrated. *If* the atom has disintegrated, the poison has been released and the cat is dead. Otherwise the atom is still intact, and the cat is alive. Since the box is closed, we have no way of knowing whether the cat is alive or dead, and the experiment invites us to consider that the superposition of the two states of the atom has 'leaked' out and affected the contents of the box, including the cat, which must also be in a superposition of being both dead *and* alive. If the Copenhagen interpretation is correct, all is in limbo until an observer looks inside the box. At this point the superposition collapses and the cat becomes either dead or alive. The absurdity lies in the fact that we experience real world cats as being either dead *or* alive, but never both.

This weird state of affairs has encouraged several other interpretations of quantum mechanics. One of these was put forward in the 1950s by Hugh Everett and is called the '*many-worlds interpretation*'. In this, there is no collapse of the wavefunction, and instead Everett posits that everything that *can* happen *does* happen. This means that when the radioactive atom and the cat enter a superposed state, physical reality splits into two separate and parallel versions: one in which the cat is alive and one in which it is dead. If this is extended to encompass all possible acts of measurement in the universe, physical reality splits into a multiplicity of separate universes, a concept that has encouraged cosmologists to propose the *multiverse*. The multiverse is a large number of parallel universes which comprise everything that exists, including all matter, all physical laws, and the fundamental constants of nature. For a fuller discussion, the reader is directed to Martin Rees's book in the Bibliography.

It has been suggested that quantum mechanics is not the final theory, but just a very good approximation to physical reality on very small scales and with very different laws that describe macroscopic bodies. The microscopic and macroscopic worlds appear to be hermetically sealed from each other. Yet it is also true that macroscopic bodies are made out of large numbers of quantum entities, and it is reasonable to ask where is the dividing line? The double-slit experiment has recently been performed using quite large molecules, buckyballs, in lieu of electrons. Beams of these C_{60} molecules (each of which weighs over a million times that of an electron) display quantum superposition and interference effects.

One aspect of the quantum world is particularly spooky. Quantum entities can pass through seemingly impenetrable barriers as if they are ghosts, a phenomenon called *quantum tunnelling*. The STM image of Figure 4 was produced by the quantum tunnelling of electrons between a fine-tipped needle probe and the carbon atoms lying under the tip. Classically, the electrons do not have enough energy to climb over the energy barrier at the tip of the probe. However, there is an alternative form of the uncertainty principle ($\Delta E \, \Delta t \geq h/4\pi$) that connects the uncertainties of another pair of complementary variables, energy (ΔE) and time (Δt). This allows a quantum particle to *borrow* the energy ΔE it needs to surmount an energy barrier, provided it pays back the loan in a time Δt governed by the uncertainty principle. In the quantum world the conservation of energy can be violated, provided it is a temporary infringement and the loan is paid back in full. This transaction allows quantum particles to spread out into classically forbidden regions and pass through barriers.

Quantum tunnelling occurs when alpha particles are emitted by radioactive nuclei, and between protons in the hot cores of stars. When two protons approach close to each other they encounter a large electrostatic repulsion, which tries to push them apart. However, their wavelike nature allows them to spread out and tunnel across the gap, bringing them close enough for the strong nuclear force to fuse them together and liberate fusion energy. Without quantum tunnelling the stars would not shine, and we would not exist.

Matter and force

In the everyday world we take for granted certain objects being *identical*. Examples are billiard balls which have the same mass, size, and composition. Billiard balls can easily be labelled, say by painting them different colours, so that as they move around the billiard table during a game, we keep track of where they go. In the quantum world, the concept of *identical or indistinguishable particles* has an altogether different and stricter meaning which forbids them being labelled. Two quantum particles are considered to be identical if the coordinates of their wavefunctions can be swapped round without changing any of the properties.

In the two-slit experiment, for example, if we try to discover through which slit the electron passes, the observation causes its quantum matter wave to get out of step with itself (a loss of coherence) and so the wavelike interference pattern on the glass screen is destroyed. The act of observation is tantamount to labelling a quantum entity, which nature forbids. The trajectories of identical quantum particles are unobservable. Unlike billiard balls, electrons can't be painted different colours.

Indistinguishable and identical particles relate to how matter and force are differentiated at a fundamental level. At the human level, we normally perceive matter and force to be different kinds of things. Matter is clearly something tangible, and forces seem to be nature's way for chunks of matter to push or pull other chunks of matter around. However, in the microworld, forces are themselves carried by particles. All the particles of the world belong to one or other of two great classes: *fermions* and *bosons*. Matter particles are fermions and force-carrying particles are bosons, which mediate forces. The two fundamental types of particles get their names from the different statistical laws that they obey when large numbers of identical particles come together. Fermions, named after the Italian physicist Enrico Fermi, obey *Fermi-Dirac statistics* and bosons, named after the Indian physicist Satyendra Nath Bose, obey *Bose-Einstein statistics*.

Symmetry plays a central role in differentiating fermions from bosons. In a system of identical particles, the probability Ψ^2 cannot change if any two of them are exchanged. On swapping a pair of particles, this leads to two possibilities: either the sign of the wavefunction changes ($\Psi \rightarrow -\Psi$), in which case the particles are fermions and have antisymmetric wavefunctions, or it is unchanged ($\Psi \rightarrow \Psi$), and the particles are bosons with symmetric wavefunctions.

Fermions and bosons each have very different properties. When fermions clump together, they avoid sharing each other's quantum

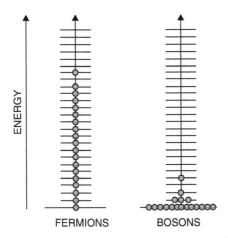

17. All particles in the world are either fermions or bosons. Matter particles are fermions and force-carrying particles are bosons. Identical quantum particles clump together on the ladder of energy levels according to their type; fermions aggregate with one particle per state, whereas bosons preferentially collapse into the lowest energy ground state.

states and instead prefer to spread themselves across the ladder of quantum energy levels, filling up the available states from the ground state upwards (Figure 17). If an extra fermion joins the crowd, it must occupy a higher rung of the energy ladder and the combined piece of matter occupies more volume. The inability of fermions to share the same quantum state prevents the electrons in an atom from getting too close to those of its neighbours, and gives rise to many properties of matter, such as its *solidity*. Bosons, on the other hand, don't care about the states occupied by the other bosons, and even prefer to occupy the same ground state level. Photons are bosons. Inside every compact disc player is a laser, in which huge numbers of bosons form a single coherent wavelike state with all the quantum waves oscillating in perfect synchrony, like a squad of well-drilled soldiers marching perfectly in step. Laser beams can be made as intense as required simply by adding more photons.

What determines whether a particle is a boson or a fermion is connected to one of the most mysterious and otherworldly of quantum properties: *spin*. Quantum particles can have an intrinsic angular momentum as if they are spinning about an axis. There is no exact counterpart to quantum spin in classical physics. The closest classical analogy to spin is rotation, for example the rotation of the Earth spinning on its axis. But here is the catch. The Earth rotates through an angle of 360° on its axis in twenty-four hours and on two successive days the Sun rises at virtually the same time. But, by how much must a quantum particle be rotated for it to look the same? Like any familiar object, a boson can be rotated 360° for this. A fermion, on the other hand, has to rotate *twice* (720°) for things to look the same. It is as if the Earth had to rotate twice, just to arrive at the next sunrise. How weird is that!

A fermion is any particle with *odd* half-integer spin (like 1/2, 3/2, etc.), while a boson has an integer spin (0, 1, 2, etc.). The spins of some bosons and fermions are given in Table 1. Fermion numbers in ordinary matter are conserved (except when annihilated by antimatter particles, which we will come across in Chapter 6). But bosons have no such limitation; they are created and destroyed in vast numbers by an action as simple as switching on and off a light.

What does it mean to say that bosons *mediate* the forces between matter particles? Imagine two skaters, on frictionless ice. As they glide towards each other, one throws a heavy ball to the other, who catches it. They both move off in new directions, each conserving momentum and energy in the exchange. A distant observer also observes the skaters but is too far away to see the ball and sees only the change in their motions. The observer would conclude that the skaters have interacted via a force. In this analogy the skaters represent fermionic matter particles and the ball a force-carrying boson. The exchange of a particle illustrates how forces operate on a microscopic scale; however,

Table 1. Properties of some bosons and fermions

Class	Name	Symbol	Mass (E/c^2, GeV)	Spin ($h/2\pi$)	Charge (e)
BOSONS	Photon	γ	0	1	0
	Weak photon	W^+, Z, W^-	80.3 (W), 91.2 (Z)	1	1, 0, −1
	Gluon	g	0	1	0
	Helium-4 atom	He-4	3.73	0	0
	Higgs particle	H^o	125	0	0
FERMIONS	Electron	e	0.0005	½	−1
	'Up' quark	u	0.002	½	+⅔
	'Down' quark	d	0.005	½	−⅓
	Neutrino	ν	About 0	½	0
	Proton	p	0.938	½	1
	Neutron	n	0.939	½	0
	Helium-3 atom	He-3	2.81	½	0

macroscopic analogies of the quantum world should always be treated with caution.

Bosons mediate the forces between matter particles by flitting in and out of existence as *virtual particles*; these are the 'heavy balls' of the microscopic world. We will see the central role they play for matter in Chapter 7. How does a virtual photon work? During its brief existence, a virtual photon borrows its energy from the vacuum, as allowed by Heisenberg's uncertainty principle. The fleeting existence of virtual particles means that even a zero-energy system can spontaneously produce energetic particles. When two electrons repel each other, the electromagnetic force between them is carried by a virtual photon, a spin-1 *vector boson*. The particle is called a vector boson because it is the quantum of a vector field, and there are three possible space directions in which the particle's spin axis can point. Quantum mechanics forbids a virtual particle ever being observed. Either the virtual quantum has to be absorbed in its entirety by the receiving particle or not at all.

The structure of the periodic table

One of the greatest triumphs of quantum mechanics is that it explains the structure of the periodic table. When Schrödinger's equation is applied to atoms, it predicts electron wavefunctions that represent various kinds of standing waves. These are called orbitals and define the regions of space where an electron is most likely to be found. An atomic orbital is specified by four quantum numbers, which define a number of different orbital shapes in three-dimensional space. The wavefunction for the ground state of the hydrogen atom is a spherical ball, with a maximum near the nucleus, gradually decreasing with radius (Figure 18). But in an excited state, the electron spends more time further away from the nucleus. Some of these higher-energy wave patterns are like hollow spherical balls, with a central core, and others are dumbbell shaped with two lobes centred on the nucleus. There are three possible dumbbell orientations, one for each direction in

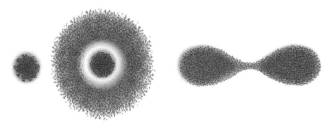

18. Spherical and dumbbell shaped electron density distributions for the hydrogen ground state (left) and some excited states. The density represents the wave pattern, and the amount of time an electron spends at a given point near the central nucleus.

space, which have the same energy. States with even higher energies have more complex electron density distributions.

Electrons fill up the atomic levels from the ground state upwards. In 1925 Wolfgang Pauli discovered the principle that governs the way electrons fill the states. If two electrons that are in the same state are exchanged, Ψ cannot change. Since electrons are fermions, this implies that $\Psi = 0$, in other words there is zero probability that two electrons occupy the same state. This is the *Pauli exclusion principle* which plays a key role in atomic structure and gives rise to the shell-like structure of the elements in the periodic table, forming groups with 2, 8, 8, 18,…, members.

In building up atoms beyond hydrogen, the electron cloud distributions are similar to the ones shown in Figure 18, but, owing to the larger nuclear charges, the electron clouds are pulled in more tightly. The next atom in the periodic table is helium, with two protons in the nucleus. The Pauli principle allows two spin-paired electrons, with opposed 'up' and 'down' spins, to go into the ground state to form the first stable *shell* in the atom. In helium, the electron clouds form a more compact ball than in hydrogen, and this is reflected in its larger ionization energy of 24.6 eV, which is almost twice as much as for hydrogen.

The next element in the table, lithium, has three protons. The extra third electron is excluded from joining the two in the first shell and instead has to go into the next available higher-energy state, and starts a new shell as a valence electron. There, it is further from the nucleus and partially shielded from the nuclear charge by the inner electron cloud. Because the outer electron is only weakly bound to the atom it is fairly easy to remove. This makes lithium chemically reactive and easily ionized, the hallmark of elements that form *metals*. Most of the elements of the periodic table are metals. Non-metals are mainly found in the upper right of a diagonal line running from boron to astatine.

A prominent feature of the periodic table is its 'eightfold' periodicity. This arises from there being *four* basic shapes of atomic orbitals: one spherical wavefunction and three dumbbells, one for each space direction. Each of these can take a *pair* of spin-opposed electrons, making eight states in a complete shell. So, in moving horizontally through the next row of the table, from lithium to neon, eight electrons are added one by one to balance the nuclear charges. Neon has a closed shell of eight electrons, which makes it a chemically inert noble gas, like helium. Beyond neon is the reactive element sodium; the added valence electron must start a new shell, and so the process repeats.

In this chapter we have looked at the profound discoveries of the early 20th century which exposed the inner structure of the atom, and the revolutionary new physics which grew up alongside: quantum mechanics. The atomic nucleus is very small, making the atom mostly empty space, inhabited by the electrostatic field that grips the electrons in their quantum shells. The nucleus contains two types of particles, protons and neutrons, which are held together by the strong short-range nuclear force. The smooth and continuous electromagnetic field envisaged by Maxwell is really made from swarms of numerous quanta. The dividing line between the quantum and the macroscopic worlds is marked by the size of a fundamental constant of nature—Planck's constant.

Heisenberg's uncertainty principle governs the behaviour of the microscopic world, and shows how particles can tunnel through barriers that are classically insurmountable. The world contains two types of particles, fermions and bosons. Fermions are matter particles, and bosons are the force-carriers. Only one fermion can occupy a given quantum state, which is the reason why fermionic matter occupies space. Bosons, on the other hand, prefer to crowd into the same quantum state where they can form coherent fields. The great success of quantum theory is that it explains the structure of the periodic table, the properties of atoms, molecules, materials, and life itself.

Although the quantum properties of matter are restricted mainly to microscopic scales, they can sometimes reveal themselves macroscopically. In Chapter 6 we will look at how aggregations of very many particles can reveal large-scale quantum effects.

Chapter 6
Quantum matter

All matter is quantum matter. The laws of quantum mechanics underpin the behaviour of molecules, atoms, and subatomic particles. Under normal conditions the quantum wavelengths of matter are too small to be discernible on macroscopic scales. But, under special circumstances, large pieces of matter containing typically 10^{23} particles can manifest large-scale macroscopic quantum behaviour. These form the focus of this chapter.

When particles condense to form a macroscopic piece of matter, two conditions have to be met for it to display large-scale quantum effects. First, the individual entities that make up the aggregate must lose their 'identities', so that their quantum interactions can extend over large distances, enabling particles to interact with other particles. Second, the particles must be freely *exchangeable*, so that the system can 'recognize' that it contains identical particles and obey the appropriate statistics: Bose-Einstein or Fermi-Dirac. The second condition means that particles need to be able to move around easily, as they can in a fluid, and the systems we are discussing are known as *quantum fluids*.

There are important differences between systems made of bosons or fermions. At high temperatures the difference is minimal, because the particles are so energetic that the probability that two are in the same quantum state is very small. But at absolute zero, the fermions must each occupy different states, in accord with the exclusion principle, whereas bosons can all drop into the lowest-energy state. Fermions fill the ladder of available quantum states up to a maximum energy, called the *Fermi level*, above which the states are empty.

The onset of quantum behaviour in a gas is illustrated in Figure 19. At high temperatures, particles move classically in straight lines, and only change their directions of motion as a result of their brief billiard-ball-like collisions. As the temperature falls, the momenta of the particles are reduced and their de Broglie wavelengths increase. At a certain characteristic temperature, the wavelengths of individual particles become comparable with their average separation, and they can interact with the others in a wavelike manner. This involves the entities being able to diffract through gaps, and interfere constructively and destructively with each other. The characteristic temperature for onset of quantum behaviour depends inversely on the masses of the particles in question, and is much larger for electrons than it is for helium atoms at the same density. If the particles are bosons, a fraction of them can condense into the *same* low-energy ground state and form what is called a *Bose–Einstein Condensate* (or BEC). As the temperature approaches absolute zero, essentially all the bosons avalanche into this state. The transition to a BEC marks a *quantum phase transition* from classical to quantum behaviour, involving coherent matter waves. A quantum phase transition therefore contrasts sharply with the classical transitions, discussed in Chapter 3, in which the solid, liquid, and gas phases arise from the competition of thermal and interatomic forces. Experiments with *two overlapping* BECs, each containing large numbers of particles, have shown that they can interfere in a wavelike manner.

Although the exotic states of matter described in this chapter involve atoms at very low temperatures, there is an example of a quantum fluid that occurs under more temperate conditions.

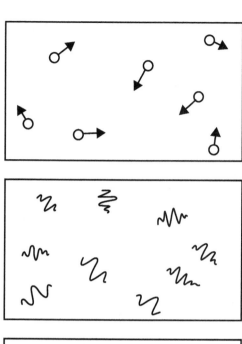

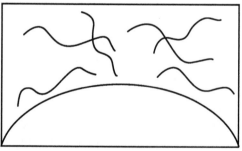

19. Atoms in a gas at different temperatures. Top: at high temperatures the atoms bounce around like little billiard balls. As the temperature decreases (centre), the de Broglie wavelengths of the atoms increase until the wavefunctions start to overlap. In this regime (bottom) the atoms lose their individual identities and quantum effects become important. If the particles are bosons, the atoms all have the same wavefunction and can settle into a Bose condensate.

It involves electrons and explains why some materials conduct electricity and others don't.

Conductors and insulators

An electric current is the movement or flow of charges between points in space, or through conducting materials. Materials that don't conduct electricity are *insulators*. Common insulators include glass, amber, many ceramics, plastics, and various crystalline solids such as diamond. The molecules in these materials are held together by strong chemical bonds, where the electrons are gripped tightly to their parent molecules and so are prevented from moving around in the material to form a current. Metals, on the other hand, contain mobile electrons and are the most common *conductors*. As we saw in Chapter 5, most of the elements in the periodic table are metals.

Metallic conduction arises naturally and spontaneously from the ease with which a valence electron can be liberated from a metal atom, resulting in a positive ion and an electron which is free to wander off into the material. A metal wire is effectively a 'highway' for free electrons. When a wire joins the terminals of a battery and so completes an electrical circuit, the electrons feel the electrical force of the battery and, being mobile, respond by flowing through the wire towards the positive terminal.

The conduction electrons in a metal form a collective fluid, which is a kind of plasma. The positive ions in a metal are immersed in the electron fluid, and the whole assembly is held together by the attraction of positive and negative charges. This is *metallic bonding*, which governs many properties of metals, not only their high electrical conductivities. The fluidity of the electron 'glue' in a metal enables the ions to assemble into close-packed crystal structures. Also, the shininess or lustre of metals comes from the inability of electromagnetic waves to pass through a region of mobile charges. This explains why light is reflected from a metal surface.

A piece of copper and a diamond are each produced by the aggregation of a huge number of copper or carbon atoms, yet their electrical conductivities differ by an enormous factor of 10^{20}. To understand why there is such a big difference, we need to consider what happens to the quantum ladder of states when a large number of atoms come together. Let's build up a solid, atom by atom. Two rungs of the quantum energy ladder are shown in Figure 20. Suppose we bring two atoms close to each other, so that their wavefunctions overlap. Because the Pauli exclusion principle forbids electrons from occupying the same energy state, the energy levels split (known as *hybridization*) to form extra close-spaced levels. As more and more atoms are added to the solid, further splitting takes place until the levels form *bands*, each containing many finely separated states in energy. There is one state in each band for each atom in the solid. Crucially, there is also a *band gap* separating the bands, in which there are no states.

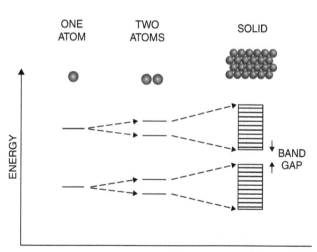

20. How electron states in a solid are *produced*. Left: two energy levels for a single atom are sharply defined; centre: for two interacting atoms the energy levels split into two; right: with many atoms the levels split further and form energy bands.

The next step in building up our material is to see how electrons populate the bands. Being fermions, the electrons obey Fermi-Dirac statistics and so exclude each other from occupying the same quantum state. Imagine pouring electrons into the solid, with one electron to a state so as to fill up the bands, from the bottom upwards. There are two possibilities. Either a band is completely filled with electrons, or it is partly full. If it is full, the electrons effectively form a 'logjam', and the solid is an *insulator* (Figure 21). In this case, the only way the material can conduct electricity is for electrons to jump across the band gap, into the next higher band where empty states are available. However, a considerable amount of energy is required to make such a big jump, and in a good insulator like diamond the band gap is 5.47 eV, which is several hundred times larger than the energy of room temperature electrons. The electrons simply can't acquire enough energy from thermal fluctuations to jump that far.

Things are very different in the *metal* of Figure 21, where the conduction electrons only partially fill a band. An electron near the Fermi level needs only to make very small energy jumps to

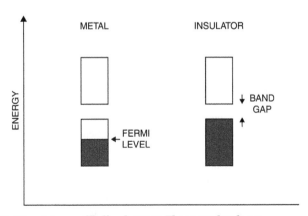

21. How states are *filled* by electrons. The energy bands are represented by boxes. Left: the partly filled band of a metal; right: the fully filled band of an insulator.

find nearby empty states where it is free to move around. A metallic conductor is therefore characterized by a partly filled band. Since each atom donates roughly one electron to the metal, the amount of space available for each electron is restricted to a volume defined approximately by the spacing of the crystal lattice. By the uncertainty principle, this spatial localization forces the conduction electrons into higher-energy quantum states, with an equivalent temperature that is of the order of 100,000 K. This is well above the boiling temperature of metals, and so the conduction electrons in metals form a quantum fluid. Matter which has a high enough density for its pressure to arise from the Pauli exclusion principle, and not from thermal motion, is called *degenerate matter*. The free electrons in a metal can be regarded as a degenerate gas, while the remainder of the electrons in it are in bound quantum states.

Although conduction electrons wander freely through the body of a metal, there is an energy cost that comes with electrical conduction. Electrons scatter from imperfections such as impurity ions, various types of crystal lattice defects, and from thermal vibrations. When they scatter, the electrons lose kinetic energy that ends up as heat. Scattering is a form of friction that results in electrical *resistance*, which is why electric fires keep you warm, and why computer microprocessor chips need to be cooled.

Superfluids and superconductors

Superfluids and superconductors display some of the most dramatic and exotic macroscopic quantum phenomena of all the states of matter. However, only very few types of atoms form quantum fluids because most materials freeze solid at temperatures well above the level where quantum effects are important, and they fail to meet the requirement for particle exchange. The best-known examples of atoms that do form quantum fluids are the noble gas isotopes, helium-4 and helium-3, which remain liquid down to absolute zero at normal pressures.

For helium-4 the characteristic temperature for the onset of quantum behaviour is 3 K, whereas for the lighter helium-3 isotope it is about 5 K. The quantum spin rule for a composite particle, such as an atom, is that if it contains an *even* number of fermions it is a boson, otherwise it is a fermion. The protons, neutrons, and electrons that make up atoms are all spin-½ fermions, and the spins always align parallel or antiparallel to each other. It follows that a system with an odd number of fermions must itself have half-integral spin. A helium-4 atom is composed of six fermions and is therefore a boson, and one of helium-3 with its five fermions is a fermion.

The study of matter at temperatures close to absolute zero began in 1908 when Dutch physicist Kamerlingh Onnes succeeded in liquefying helium at a temperature of 4 K. By pumping on the liquid to lower the pressure, even lower temperatures could be attained, making it some of the coldest matter in the universe, with temperatures below even that of the cosmic microwave background temperature of 2.7 K.

Normal fluids are *viscous*. Viscosity is the internal friction or 'stickiness' between the molecules in a fluid that tends to resist uniform flow. It takes energy to overcome viscosity; think of swimming in a pool. In 1937, Russian physicist Pyotr Kapitsa discovered that when helium-4 is cooled below a temperature of about 2 K, the liquid spontaneously makes a phase change to an exotic *superfluid* phase in which the viscosity dropped precipitously by a factor of more than 10^{11}. Superfluid helium-4 has a range of dramatic quantum mechanical properties, such as the ability to flow though very narrow capillary tubes without *any* friction, and form a thin film (called a *Rollin film*), which can creep *up* the sides of a containing vessel and climb out (Figure 22). When it is heated, a superfluid produces a spectacular spouting fountain (Figure 23). Below the quantum characteristic temperature, all the bosonic atoms of helium-4 tend to condense into lowest energy quantum state, and form a Bose Condensate,

22. The creep of superfluid helium over the sides of its container to drip out underneath.

23. The fountain effect. Mildly heated superfluid helium produces a spouting fountain.

in which the overlapping matter waves form a single coherent quantum entity where billions of atoms act as if they were one giant 'atom' moving collectively. This is analogous to a battalion of perfectly drilled soldiers marching in step. If one moves, the others must follow.

The quantum coherence in superfluids explains why superfluid helium can flow frictionlessly through fine capillary tubes. Normal viscous liquids cannot flow through such narrow channels, since any slight irregularities or roughness on the wall scatters fluid, creating viscous forces large enough to block the flow. But in a superfluid, the viscous forces are rendered ineffective because every atom must obediently follow all the others, and macroscopic flow becomes possible. Collective behaviour on a macroscopic scale is also found in a *laser*, in which a Bose condensate of spin-1 photons occupies a single quantum state, in which there is no limit to the number of particles. A laser beam is a coherent light wave and a Bose condensate of helium-4 is a coherent matter wave.

After Onnes had discovered how to liquefy helium, he used the technique to cool metals down to very low temperature to study their electrical conductivities. In 1911 he was astounded to discover that the resistance of mercury suddenly fell precipitously to zero below a transition temperature of about 4 K. This was the discovery of *superconductivity*. The current (or supercurrent) that flows in a superconductor does not decay measurably over timescales of years. Other metallic superconductors were soon discovered. A key defining property of a superconductor is that it completely expels all magnetic fields from its interior, except in a very thin surface layer. A superconductor does this by producing internal currents that generate magnetic fields to cancel the externally applied ones. This is known as the *Meissner effect*, and it can have spectacular consequences. If a bar magnet is lowered into a superconducting bowl, the work done in expelling the

magnetic flux can support the weight of the magnet, which is
levitated and floats above the bowl.

Superconductivity was eventually understood to be a rather subtle
effect. In 1957 John Bardeen, Leon Cooper, and Robert Schrieffer
(collectively known as BCS) deduced that paired-up electrons
(called *Cooper pairs*) carry the supercurrent. In the normal
non-superconducting state two electrons repel each other. But in
low-temperature superconductors, the Cooper pairs resemble
bosons that interact dynamically with small distortions in the ion
lattice, to form a weak collective bound state. A Cooper pair has
been pictured as being something like a two-electron 'molecule'
within the superconductor, an entity that can exist over distances
of many millions of atomic spacings. The theory describes Cooper
pairs condensing into a single macroscopic quantum condensate,
a *superfluid*, and moving as a single unit. This is reminiscent of
the BEC phase in helium-4; but in this case the fluid contains
charged particles and therefore can be manipulated with electric
fields. Provided the Cooper pairs remain intact, the superfluid
flows through in the superconductor with no electrical dissipation
or resistance.

The weak binding of Cooper pairs makes the conventional
low-temperature superconducting state a fragile one that can easily
be destroyed by thermal fluctuations and such materials generally
have transition temperatures below about 20 K. But in 1986,
Georg Bednorz and Alex Mueller made a surprise discovery of a
remarkable new class of ceramic superconductors, with much higher
superconducting transition temperatures. These high-temperature
superconductors are currently not understood and, to date, the
highest temperature superconductor known is a relatively common
compound, hydrogen sulphide (H_2S), which, under very high
pressures, is a superconductor at 203 K ($-70°$ C). The ultimate goal
is to discover a room temperature superconductor. Such a material
could, for example, open the possibility of transmitting electrical
power very efficiently over large distances.

The Josephson junction

If two metals are separated by a very thin insulating barrier, no more than a few hundred atoms thick, they can form what is known as a *tunnel junction*, which can support a small electron current flowing across by quantum mechanical tunnelling. In 1962, English physicist Brian Josephson realized that if the metals were replaced by *superconductors*, and formed a quantum system, a larger supercurrent can flow between them. In Chapter 5 we saw how the square of the amplitude of the wavefunction is interpreted as a probability and, for a single particle, the phase of the wave is largely irrelevant. But this is no longer true when there is a *phase difference* between two interacting quantum objects. In a *Josephson junction* the phases of the wavefunctions differ across the barrier, and the resulting supercurrent flowing between them turns out to be related to the *gradient* of the phase of the wavefunction.

The Josephson junction has interesting properties. First, since the Cooper electron pairs in the superconductors carry charge, the junction can be manipulated by applying external electromagnetic fields. If a voltage is applied to the junction, it bursts into oscillation (typically at microwave frequencies for applied voltages in the millivolt range). This is because the phases of the wavefunctions across the junction 'beat' with each other rather like two out-of-tune piano strings, which drives an alternating current, the Josephson current. The reverse is also possible. If the junction is exposed to an oscillating microwave field, the supercurrent shows quantized 'steps' that occur at particular voltages. These quantized steps provide a very precise way of measuring voltages (to 1 part in 10^8), or alternatively of measuring a fundamental constant of nature, e/h, with great precision. The degree of quantization is so precise that Josephson junctions are used in metrology to define the volt.

There are other possibilities. Two Josephson junctions can be connected in parallel, to form what is known as a SQUID (or *Superconducting Quantum Interference Device*). In the SQUID circuit, when a magnetic field is applied, the phase relationship between the two junctions is altered, and the SQUID can be used as a magnetometer to measure very small magnetic fields, such as those produced by tiny currents in the brain. SQUIDs can also be used as high-speed switches being developed for quantum computers.

Shapes of quantum matter

Understanding the workings of a complex physical system in 3D presents a formidable challenge, and to make progress it's often useful to study a simplified system that retains enough of the features of the original one to be relevant. An example of this is the reduction of a 3D system to 2D, a motivation that in 1980 led Klaus von Klitzing to the Nobel-Prize-winning discovery of the *Quantum Hall Effect* (QHE).

The classical *Hall effect* had been discovered a century earlier by the American physicist Edwin Hall when he was studying electric currents flowing in a thin metal sheet. He discovered that when a magnetic field is applied perpendicular to the sheet, a voltage called the *Hall voltage* appears across the sample. When any charged particle, such as an electron, tries to move in a straight line through a magnetic field, the field tends to bend it into a circular orbit. In Hall's experiment, the presence of the Hall voltage could be explained by the effect of the magnetic field bending electrons sideways, towards the edges of the strip where electrical charge builds up. The Hall voltage, and its associated electrical resistance, change smoothly as the magnetic field strength is varied.

Von Klitzing was interested in the quantum analogue of the Hall effect, and had set up an experiment to study currents flowing in a thin 2D layer of electrons, cooled to very low temperatures so that the electrons formed a coherent quantum system. The layer was immersed in a very strong perpendicular magnetic field, so that the electrons could complete fully circular orbits, lying in the plane of the sample. This circular motion is reminiscent of electron motion in the Bohr model of the atom; and we can think of the 2D electron layer as being divided up into a number of regions defined by the sizes of these closed orbits. When this circular motion is quantized, a ladder of discrete quantum energy states forms, which produces an energy gap separating occupied and empty bands, just as in an ordinary insulator.

When von Klitzing varied the strength of the magnetic field, he discovered a dramatic effect: the Hall resistance did not change smoothly, but jumped discontinuously between steps on a 'staircase' of broad plateaux, revealing quantization on a macroscopic scale (Figure 24). This is how the Quantum Hall Effect was discovered. A second big surprise was the discovery that the Hall conductance (the inverse of resistance) is quantized in exact integer multiples of a fundamental constant of nature: e^2/h. Subsequent measurements have shown that the level of quantization holds to an extremely high precision of at least 1 part

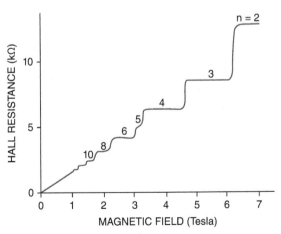

24. The integer Quantum Hall Effect: the measured Hall resistance is plotted against the strength of the applied magnetic field. The integers n refer to different topologies of the electron wavefunctions.

in a billion. This is found to be robustly insensitive to the details of the sample geometry and imperfections arising from its preparation, in sharp contrast with the classical Hall effect, where, depending on conditions, there is considerable variability.

This unprecedented behaviour could not be understood on the basis of the standard theory of electrical conduction. While the bulk of the sample behaves like a standard band gap insulator, something unusual happens at the edge. There, an electron cannot complete a circular orbit without hitting the 'hard' outer edge, bouncing back, and skipping along the edge in a series of jumps. These unidirectional orbits are known as '*skipping orbits*' which form an edge current. A QHE material has the interesting and unusual property that it is an insulator in its bulk, and a conductor on its surface.

The Quantum Hall system has come to be understood by recognizing that the electron layer must be considered as a macroscopic quantum system in its own right. The special properties of the system relate to the various shapes, or topologies, of the electron wavefunctions. Topology is a branch of mathematics dealing with the geometrical properties of objects which are unaffected by continuous deformations such as bending, squashing, or stretching. Objects belong to different *topological classes* if they are pierced, or parts of them are glued together, to make holes. Consider the examples of a mug, a bagel, and a pretzel shown in Figure 25.

A mug (let's say made out of soft clay) can be deformed into a bagel shape and back again simply by a continuous squeezing and pulling process without the need to cut any holes. The mug and the bagel are therefore topologically equivalent; they have one hole each and belong to the same topological class. However, neither of these shapes can be deformed to the pretzel shape without piercing two extra holes; the pretzel therefore belongs to a different topological class.

25. A coffee mug is topologically equivalent to a bagel, because they both have one hole; but they both belong to a different topological class from the pretzel with its three holes.

In the Quantum Hall system, each plateau in the measured Hall conductance (Figure 24) represents a different topological class, each of which is labelled by an integer, n. At a very crude level the different classes can be thought about as different ways of 'knotting' the electron wavefunction. Topology has added a new method to classify matter. For example, it is possible to think of deforming one type of solid insulator into a different one, by various actions. These might include physical deformations such as stretching, compressing, or bending the material. These actions alter the energy levels and bands in the solid. However, provided the insulator *always* remains an insulator during a particular deformation process (in other words there is always a forbidden energy gap present), the final insulator must belong to the same topological class as the initial one. But if at any point the energy gap should close up, causing the material to become, even briefly, a metal, the initial and final insulators belong to different topological classes.

Since the discovery of the Quantum Hall Effect, our understanding of what are now called *topological insulators* has advanced considerably. Moving away from two dimensions, to the world of practical materials, topological insulators in 3D have now been discovered. It turns out that many common insulating chemical compounds, for example those containing elements such as bismuth, selenium, and tellurium, are topological insulators, all characterized by conducting surfaces and insulating interiors. These materials display remarkable properties; if a topological insulator is cut into pieces, new conducting surfaces appear where cuts have been made. In 2016, David Thouless, Duncan Haldane, and Michael Kosterlitz were awarded the Nobel Prize for their contributions to understanding these new exotic states of matter.

The kilogram

According to Isaac Newton, a key property of matter is mass, and the *kilogram* is the fundamental unit of mass in the SI system.

It is also the one remaining unit to be defined by a physical object, a lump of platinum–iridium alloy cast in 1879 and kept in a secure vault in the International Bureau of Weights and Measures (BIPM) in Paris. The object is affectionately known as '*Le Grand K*' (Figure 26). Once every forty years, it is taken out so that replica kilos from around the world can be compared with it.

The problem is that 'Le Grand K' is not stable: its mass has diverged from those of its clones. The difference is small and amounts to a

26. **The International Prototype Kilogram is stored inside three bell jars in a safe in a basement in the Paris suburb of Sèvres.**

mass loss of just 50 micrograms a century, which is about the weight of a grain of sand. A weight loss of two parts in 10^7 per century may appear to be small, but it is nevertheless significant. In 2018 it was therefore decided that the physical kilogram will be replaced by an 'electrical' kilogram that relies not on the integrity of a physical lump of matter, but on the measurement of fundamental physical constants. Matter will be weighed against the electromagnetic force.

The new and more precise definition of the electrical kilogram relies on the Josephson and Quantum Hall effects, which provide the high precision of the electrical voltage and current measurements needed. The Josephson junction standard will be used as a voltage standard and a Quantum Hall effect device for the current/resistance standard. With the proposed new electrical standard kilogram, laboratories anywhere in the world will be able to define the kilogram *in situ*, without the need to transport a material object. The net uncertainty in the weight of the electrical kilogram is 1 part in 10^8.

This chapter has explored some examples of quantum fluids, matter that displays quantum effects on macroscopic scales. Quantum fluid behaviour explains the difference between conductors and insulators, as well as the dramatic properties of superfluids and superconductors. These exotic macroscopic quantum systems also have practical uses, and have led to new and very precise ways to make electrical measurements. These techniques will be used to make a greatly improved definition of the fundamental unit of mass, the kilogram, one which is linked to fundamental physical constants, and not to a physical piece of matter.

In Chapter 7 we will zoom in on subatomic particles and look at the smallest and most fundamental particles of matter.

Chapter 7
Fundamental particles

All the composite forms of matter—nucleons, atoms, molecules, living creatures, planets, stars, and so on—are built out of a small number of different types of particles which interact in different ways via the forces of nature. It is the forces that give rise to the enormous variety and many forms of matter. When discussing matter, we must therefore include the forces. There are four fundamental forces: the two long-range forces of gravity and electromagnetism, and the two short-range forces that operate in the nucleus, the strong nuclear force, and the weak nuclear force. The strong force binds nuclei together, and the weak force is connected with certain types of radioactivity, and the energy producing processes in the stars.

Just over a century ago, it was reasonable to have imagined that the ultimate building blocks of the universe were atoms. By 1932, the atoms of normal matter could be explained in terms of just three types of subatomic particles, protons, neutrons, and electrons. This beautifully simple picture of the fundamental components of matter did not last long. By then, matter was starting to be probed at higher energies, and theoretical developments in quantum mechanics had taken place which would radically change our understanding of the microscopic world.

While the Schrödinger equation successfully describes much of the subatomic world, it does not describe the properties of particles moving at speeds close to the speed of light. This shortcoming was significant because, for example, while the speed of an electron in the ground state of a hydrogen atom is only about 1/137th of the speed of light and therefore non-relativistic, this is not true for the innermost electrons in heavy atoms like gold, in which speeds approach half the speed of light. (The number 1/137.036 ... is a constant of nature known as the fine structure constant.) So, in 1928 English physicist Paul Dirac set out to make quantum theory consistent with special relativity.

The first attempts to unify quantum theory with special relativity had foundered on a mathematical problem. In classical wave equations, space and time are treated separately, but in special relativity, they are woven into a single fabric: 4D spacetime. Dirac discovered that he could incorporate special relativity by using matrices to divide the equation into four parts, resulting in the famous *Dirac equation*. The property of electron spin came out naturally from the theory, directly from special relativity. The electron also emerged as a fundamental particle of zero size. Two of the four parts of the equation were easy to interpret: these correspond to the electron spinning clockwise (up) or anticlockwise (down). But the puzzle was: to what did the other two correspond?

Antimatter

In special relativity, the energy E of a system appears in the equations as a squared quantity, i.e. as E^2. Since a square root can have positive *and* negative values, twice as many solutions result. The negative value of the square root implied that an electron has *negative energy*—something that made no sense. Dirac realized that a negatively charged electron with negative energy could be interpreted as a positively charged electron with positive energy.

In effect he transferred the negative sign of the energy to the electrical charge where it converted a negative electron into a positive electron, or *positron*, a particle of *antimatter*. Matter and antimatter are mirror images of one another; whatever one does, the other does the opposite. For example, if an electron spins clockwise, a positron spins anticlockwise.

Did the positron really exist, or was it just a quirk of the mathematics? A few years after Dirac's prediction, the physicist Carl Anderson was studying cosmic ray electrons. Cosmic rays are energetic particles that originate in the Sun and outer space and bombard the Earth. Anderson had set up a cloud chamber detector inside a strong magnet to record the tracks of the cosmic rays. When a charged particle passes through a cloud chamber, it produces a condensation trail of tiny liquid droplets behind it, like the contrail of a jet aircraft flying in the stratosphere. When a negatively charged electron passes through a magnetic field it is deflected sideways, always in the same direction. Anderson observed that in addition to the electrons, there were particles with the same mass as the electron, but with positive charges, which were deflected in the opposite direction. Dirac's positrons had been found.

When an electron and a positron meet, they annihilate each other, and convert all of their rest-mass energy, a total of 1.022 MeV, into a burst of gamma radiation. The process is symmetric. If an energetic gamma ray photon collides with an atom, the strong electric field around the atomic nucleus causes *matter* to be converted directly from *energy* in a process called *pair production* (Figure 27). The electron–positron pairs Anderson observed were pair production events.

All particles, not just electrons, have antiparticle partners, and it was not long before other types were discovered, always one of a pair, where electrical charge is strictly conserved in equal and

27. Pair production: matter and antimatter are created directly from the energy of a photon. A gamma ray photon enters from the top, passes close to an atomic nucleus, and produces an electron (left-handed spiral), and a positron (right-handed spiral). A recoiling atomic electron emerges downwards.

opposite amounts. If a negatively charged antiproton passes near a proton, the charges cancel so as to leave behind two neutral particles, a neutron, and an *antineutron*. Whole anti-atoms have been made in which a positron is combined with an antiproton to make *antihydrogen*. The light from a quantum transition in antihydrogen was observed for the first time in 2017 and was found to have a wavelength that is identical to that of the corresponding transition in normal hydrogen.

Quantum fields and forces

In early quantum theory, the particle took centre stage, and the theory provided the scaffolding needed to calculate the particle's quantum states and its associated energy levels. This approach however could not explain how particles were created and destroyed. The next step was to develop quantum mechanics into a more general framework: *Quantum Field Theory* (QFT). QFT places the emphasis on the field, and assumes that space is filled by a number of interacting particle and force fields, each of which is characterized by its quantum states. The primary concept here is that of an all-pervasive quantum field, in which the particles (the field quanta) are fluctuations. Just as photons are the field quanta of a quantized electromagnetic field, electrons are the field quanta of a quantized electron quantum field.

The first successful quantum field theory to be constructed was quantum electrodynamics (QED), which describes the interaction of light and matter. At the start, it was beset by theoretical difficulties. One of the most glaring was the interaction of an electron with its own electromagnetic field, the so-called self-energy of the electron. The Coulomb electric force from an electron (the source of the field) varies as the inverse of the distance squared. So, if the electron has zero size, the force ought to be infinitely large at the position of the particle (one divided by zero distance is infinity). The energy in the electromagnetic field surrounding an electron should therefore be infinite and, since energy and mass are equivalent, the mass of the electron should also be infinite.

The irksome infinities were related to the electron's self-energy and the polarization of the vacuum. To visualize vacuum polarization, imagine trying to create the most perfect of all vacuums, by removing all particles from some region of space. When you have done that, Heisenberg's uncertainty principle tells us that *something* still remains. Quantum mechanics allows virtual particles to pop in and out of existence constantly; the vacuum has a finite energy density. So-called empty space is really a seething cauldron of activity, filled with an energy called the *vacuum energy*. The electric field surrounding an electron contains a bubbling sea of virtual particles. The self-energy of the electron arises from the interaction of its intrinsic, or 'naked', charge with these particles and so produces an extra energy density. This interaction confers upon the electron something called its *electromagnetic mass*.

The theorist Hans Bethe circumvented the problem of infinities with a mathematical trick, called *renormalization*, which involves subtracting one infinite quantity from another to leave a finite answer. Seen by some to be unsatisfactory, renormalization was found to work well in practice and enabled QED to be developed into the most accurate physical theory of which we know. One of the first tests it faced was to explain something called the *Lamb shift*. Willis Lamb had observed that a particular quantum state of hydrogen was split into two close-spaced energy levels. The Lamb shift is caused by the self-interaction of the electron with its field and could not be explained by Dirac's theory. The splitting is predicted extremely precisely in the modern theory of QED, developed in 1947 by Richard Feynman, Julian Schwinger, and Shin'ichiro Tomonaga.

Physicists had also been studying the strong and weak nuclear forces. Nuclear forces differ from the electromagnetic and gravitational forces in that they operate over very short ranges, below around 10^{-15} metres. On scales of the size of an atom (10^{-10} metres) nuclear forces essentially don't exist. Nuclear forces are like Velcro: when two pieces of Velcro are in contact they are strongly attached, but when they are pulled apart they feel no force whatsoever. The strong force is also indifferent to electrical charge; it binds the positively charged protons and neutrons in the nucleus together equally strongly. Recall from

Chapter 5 the analogy of the two ice skaters exchanging a heavy ball to illustrate the microscopic nature of forces. In any short-range force, the mediating force particle must have a large mass and a short lifetime, as required by the energy-time form of the uncertainty principle. In 1935, Hideki Yukawa proposed that the strong force involves the exchange of a virtual particle called a *pion*. The mass of a pion lies in between those of an electron and a proton, and belongs to a class of particle called a *meson* (from the Greek word meaning 'intermediate'). Pions are unstable and live brief lives inside nuclei; but when a nucleus is struck very hard by another particle, pions can be driven out into the world as real particles with short lifetimes. These short-lived pions were observed in 1947 in cosmic-ray particle interactions. The pion model for the strong force was an approximate theory, and has come to be replaced by the quark theory, which is described below.

The weak nuclear force, or more properly the weak interaction, has a more subtle character. It is responsible for beta radioactivity and also plays a key role in how the Sun burns. A lone neutron is unstable and will decay into a proton and an electron in around ten minutes. Inside the nucleus however the neutron is more stable, but if the atom is radioactive a neutron can decay into a proton, changing the chemical identity of an atom to the next higher element in the periodic table. This *beta-decay* process is accompanied by the emission of an energetic electron, the beta particle of radioactivity. When beta-decay was first studied, it was found that electrons were 'spat out' of nuclei with a continuous range of energies, from zero up to a maximum value. This was problematic, because it appeared that the laws of energy and momentum conservation were being violated.

Wolfgang Pauli hit on the right solution when he suggested that an unseen and mysterious particle, a *neutrino* (or 'little neutron') was emitted along with the electron, so as to balance energy and momentum. In beta-decay the emitted particle is an antineutrino; in the inverse process, inverse beta-decay, protons are converted into neutrons by emitting positrons and neutrinos. Neutrinos have very low mass, no electrical charge, and can penetrate very deeply into matter. The Sun and the stars produce trillions of them every second, but they interact so weakly with ordinary matter that a neutrino can pass through a light year of lead (ten trillion kilometres), and still have only a 50:50 chance of hitting a nucleus. We are blissfully unaware of the vast numbers of neutrinos that pass unfelt through our bodies. Each second a billion neutrinos zip through your thumbnail, day and night. The elusive nature of the neutrino meant that it was not detected until 1957, in reactions in a nuclear reactor.

From the description of the weak interaction just given it is reasonable to ask: why is it regarded as one of the four fundamental forces of nature? In their quest to understand a deeper physical reality, physicists often seek apparently different aspects of nature that have a common cause and so are amenable to unification. One example was Maxwell's unification of the electric and magnetic fields, which as we have seen revealed the more fundamental electromagnetic field. Another was Dirac's unification of special relativity with quantum mechanics, which resulted in the prediction of antimatter. A third example is the unification of the electromagnetic and weak nuclear forces (the *electroweak* interaction), which was postulated in 1967 by Sheldon Glashow, Steven Weinberg, and Abdus Salam. In essence, their theory proposed that the particles carrying forces in electromagnetism and the weak nuclear force are really the same. They only appear to be different because the force-carrying bosons (the W and Z particles) have mass in the weak interaction, but in electromagnetism the corresponding particle, the photon, is massless.

The weak nuclear force is only about 1/1000th as strong as the electromagnetic force, yet the two have much in common. To understand how they relate to each other, recall that the long-range electromagnetic force is carried by the photon, which is massless and carries no electrical charge. This can be compared with the W and Z carriers of the weak force, which have masses about 100 times the proton mass, about the same mass as a silver atom. W particles can be positively or negatively charged, and the Z is neutral. When a neutron undergoes radioactive beta-decay, it emits a charged boson, the negative W particle. In the interaction, the neutron recoils to conserve energy and momentum, and so the W has the key property of a *force carrier*. The properties of the four fundamental forces are summarized in Table 2.

Unlike the other forces of nature, the W changes the *identity* of a particle. When a neutron converts into a proton, one unit of negative charge is removed, and the emitted W subsequently decays into an electron and a neutrino. The short lifetime of the W is related to its large mass; it is a heavy boson with a limited range. In electromagnetic interactions, a virtual photon carries no charge, and strictly speaking should be regarded as forming a *neutral current*. The analogous neutral current in the weak interaction is carried by a neutral boson, the Z particle. The three bosons (W^+, W^-, and Z^0) are called *intermediate vector bosons*. In 1983, experiments at the *Super Proton Synchrotron* accelerator at CERN in Geneva directly observed the Ws and the Z^0.

But there was still something missing—an understanding of the origin of mass. Quantum field theory predicted that the W and Z particles should, like the photon, be massless. So how do the Ws and Zs get their large masses? A major clue came from an idea in the theory of condensed matter—the behaviour of photons in superconductors. In Chapter 6 we saw that a superconductor is highly intolerant of external magnetic fields (the Meissner effect), and tries to expel them completely from its bulk by generating supercurrents that cancel them out. This property of a superconductor applies to all magnetic fields, including those associated with the quantum fluctuations of virtual photons. When a virtual photon flits into existence inside a superconductor, the superconductor responds by generating supercurrents that try

Table 2. Properties of the four fundamental forces

Force	Quantum (source charge)	Range (metres)	Mass (E/c^2, GeV)	Relative strength	Examples in nature
Strong	Gluon (colour)	10^{-15}	0	1	Hadrons (quarks)
Electromagnetic	Photon (electric)	∞	0	1/100	Binds electrons to nuclei in atoms
Weak	W, Z	10^{-18}	0.08 (W), 0.09 (Z)	1/100,000	Beta radioactivity
Gravity	Graviton (mass)	∞	0	10^{-40}	Solar system, galaxies

to cancel out the fluctuating magnetic field of the photon. As a result, the field of the photon is weakened, and more energy is needed to sustain the fluctuations. This extra energy confers mass on the virtual photon, called the *effective mass*. Photons *acquire* mass by being inside a superconductor.

This profound idea was seized upon to explain the origin of the masses of the W and Z bosons. Could it be that all of space is filled with an unidentified quantum field, which, by analogy with superconductivity, interacts with fundamental particles to give them their masses? In 1964 physicists Peter Higgs, Robert Brout, and François Englert proposed such a quantum field, the *Higgs field*, which fills the universe. If true, it implied that we would, in effect, be living inside a cosmic superconductor. In this theory, all fundamental particles are massless until they interact with the Higgs field by what is called the *Higgs mechanism*. A crude physical picture of this is that particles acquire mass by 'sticking' to the field as if they are trying to move through thick treacle. The Higgs field contributes a finite energy density to the vacuum and, because it has no preferred direction in space, is a scalar field.

The Higgs mechanism relates to the unification of electromagnetism and the weak interaction via spontaneous symmetry breaking. Recall from Chapter 3 how symmetry is broken when a liquid freezes, or when the temperature of a piece of iron falls below the Curie temperature and becomes a magnet. At very *high* temperatures and energies, the W and Z particles don't interact with the Higgs field, rendering them massless, like the photon. But at *low* temperatures the symmetry breaks, and the Ws and Zs interact with the Higgs field, from which they acquire their masses. In the electroweak interaction, the terms 'high' and 'low' energy relate to a critical energy of about 100 GeV, which defines the scale of the electroweak force.

The quantum particle of the Higgs field is the *Higgs boson*. The search for this much sought-after particle was a key objective of experiments that started in 2013 in CERN's *Large Hadron Collider* (the LHC). The experiments involved colliding beams of protons with each other with a combined energy of 8 trillion electron volts (8 TeV, or 8×10^{12} eV). These collisions hit the Higgs field very hard, and made it oscillate and produce a quantum of field excitation: the Higgs boson. The measured mass of the particle (125 GeV/c^2) was consistent with its very short lifetime of about 10^{-22} sec. The Higgs particle was not observed directly, but its existence was inferred unambiguously from the particles into which it subsequently decayed.

Quarks

To study the smallest particles of matter a very powerful microscope is needed. If the objects of study are smaller than a nucleus they must be probed with very short-wavelength, high-energy particles. High-energy particle accelerators are the microscopes of the subatomic world; the bigger and more powerful the accelerator, the stronger the microscope.

From the 1940s to the 1960s, many hundreds of new types of subatomic particles turned up in accelerator experiments. In these, high-energy protons were smashed into targets, producing a plethora of unstable particles. The situation was not unlike that of the previous century, when chemists had been confronted with a bewilderingly large number of chemical elements with different properties. Trying to make sense out of all the particles was, as Richard Feynman put it, 'like trying to figure out a pocket watch by smashing two of them together and watching the pieces fly out'. By 1954, so many particles had been discovered that Fermi complained, 'If I could remember the names of all these particles, I would have been a botanist.'

Subatomic particles belong to two broad families: *hadrons* and *leptons*. Hadrons (from the Greek word meaning 'thick') have a family resemblance to protons and neutrons in that they feel the strong nuclear force, the weak interaction, and electromagnetic forces. The lightest members are protons and neutrons, but most hadrons are much more massive, and are in excited states with high internal energies and short lifetimes. The hadron family is subdivided into two further types, *baryons* and *mesons*. The word baryon means 'heavy'. Leptons (from the Greek for 'thin', or 'small') are particles like electrons and neutrinos, and feel only the weak and electromagnetic forces.

Physicists scrutinized the mass of data in the hadron zoo, searching for patterns that might reveal a classification that would be for the subatomic world as profound as Mendeleev's periodic table had been for the elements. In 1964 Murray Gell-Mann and George Zweig did indeed find patterns among the plethora of hadrons, and concluded that all the particles could be built out of *quarks*, elementary particles with fractional charges. Gell-Mann chose the name 'quark' from a line in James Joyce's book *Finnegans Wake*: 'Three quarks for muster Mark.'

Normal matter is made from two types of quarks: an 'up' quark (u, mass 2.3 MeV/c^2), and a 'down' quark (d, mass 4.8 MeV/c^2). Each type carries a different fractional electrical charge; the up-quark carries +⅔ of the electron charge, and the down quark –⅓ of it. Baryons are made from three quarks, and mesons from a quark–antiquark pair. A proton (Figure 28) can be imagined as a tiny sphere, about 10^{-15} of a metre across, containing three quarks: (u, u, d) giving it a net charge of +1 unit. The neutron is similar, but has (u, d, d) quarks giving it zero charge. The quarks move round inside a nucleon at almost the speed of light, bouncing off each other and off the walls. All the hadrons in the particle zoo could seemingly be explained as various bound states of quarks. Was this just a clever way of organizing the hadron zoo, or did quarks really exist?

Decisive evidence that quarks really do exist came from a series of *'deep electron scattering'* experiments made in the 1960s and 1970s by Jerome Friedman, Henry Kendall, and Richard Taylor.

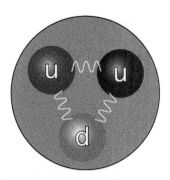

28. The lightest baryon, the proton, is made from three quarks, two 'up' ('blue' and 'red' colour charges) and one 'down' ('green' colour charge). All three colours (shown here by differences in shading) must be present, making the particle colourless.

They used the 3-kilometre-long *Stanford Linear Accelerator* (SLAC) to bombard protons with electrons having speeds close to the speed of light. The electrons were seen to scatter from tiny granular objects moving about very fast inside the proton. These tiny objects were the sought-after quarks.

Soon, other more massive varieties of quarks were discovered, coming in different 'flavours' that signify their symmetry properties. There is a *charm quark* (*c*) with a mass of 1.3 GeV/c^2, the *strange quark* (*s*) with mass 0.95 GeV/c^2, and the *bottom quark* (*b*) with mass 4.2 GeV/c^2. In 1995 an even more massive 180 GeV/c^2 *top quark* (*t*) was discovered using the high-energy collider at Fermilab. The unstable heavier quarks decay rapidly to the lower mass states. There are six types of quarks, and all are spin-½ fermions. As far as we know, quarks are the ultimate fundamental particles of matter.

Quantum chromodynamics (QCD)

A very strong force is required to bind three very energetic quarks together in a hadron. The force particles that tie quarks together are called *gluons*. But how is it that *three* spin-½ fermions are able to coexist in a nucleon, apparently in violation of the Pauli exclusion principle? It turns out that quarks and gluons carry a different kind of charge called *colour charge*, which is a measure of the strength with which the particles interact via the strong force. Unlike electromagnetic charge, which comes in two varieties, positive and negative, colour charge comes in three varieties: red, green, and blue. The associated *colour force* is the basic force of the strong interaction. As in electromagnetism, colour charge is a strictly conserved quantity. It is important to appreciate that the colour names are just labels, and do not signify anything that we would recognize in the familiar world as colours. The labels merely help us distinguish between different quantum states and recognize that, say, a blue up-quark is different from a green up-quark. This may sound a little contorted, but the key quantum rule is that quarks only form combinations that are *colourless*. In a hadron, all three colours must be present. Mesons are automatically colourless because colour and anticolour charges cancel out in quark–antiquark pairs. To add another layer of complication, it turns out that quarks can also change colour. Gluons are themselves coloured and so can transfer colour charge from one quark to another.

One of the most striking properties of quarks is that they are never seen alone. The '*principle of confinement*' relates to the way quarks interact with the gluon fields inside a hadron and keeps them tightly bound inside. The concept of 'tying' things together is particularly apt for quarks because the force law between quarks behaves as if they are connected by elastic strings. If you stretch an elastic band, the force needed increases the more it is stretched. But if no force is applied, the elastic is slack. The force law between quarks works in precisely the opposite way to the inverse square law of gravity or electromagnetism: *the force between quarks increases the further apart they get*!

The interactions between colour charges are described in a quantum field theory, *quantum chromodynamics*, or QCD, constructed by David Gross, Frank Wilczek, and David Politzer. QCD differs from QED in a key respect. In electromagnetism, two uncharged photons can pass freely through each other, oblivious to each other's existence. The airwaves are full of electromagnetic signals from multiple radio and TV stations and mobile phone signals, all of which cross and superimpose without affecting each other. But gluons carry colour charges, which means they interact strongly with each other, a property that has huge consequences. If you hit a quark inside a nucleon hard enough it can emerge a small distance into the outer world, but it is still connected to the other two quarks via the colour force, which as we have said behaves like a piece of strong elastic. The further out the quark emerges, the stronger the force it feels pulling it back, and, to pull it out even more, more energy is needed. Eventually so much energy has to be provided that the elastic snaps. When that happens all the energy stored in the 'stretched elastic' of the colour field converts into matter and antimatter, and a quark–antiquark pair forms. Strings of quark–antiquark pairs are commonly produced in particle accelerators experiments, and are known as *particle jets*.

The Standard Model

The Standard Model of Particle Physics (Figure 29) is one of the very great achievements of science, and is the distillation of many decades of work by enormous numbers of scientists. The model is well validated by experiment, and brings together all the fundamental matter and force particles we know of. The foundations on which the standard model is built are QED, the electroweak interaction, and QCD. All the fundamental particles needed to build normal matter are here: from quarks, to nuclei, to atoms, to chemistry, to life, all the way up to the visible universe. Gravity is not included in the model because there is at present no viable theory of quantum gravity.

The matter particles, the fermions, are shown on the left, and the force particles, the bosons, are on the right, with the Higgs

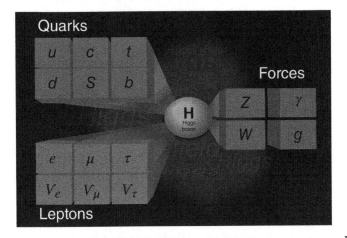

29. **The Standard Model of particle physics, the sum of our knowledge of the fundamental matter and force particles from which the universe is made.**

boson in the centre. Each fermion has an antimatter twin, which are not shown here. All of the matter of the everyday world is built from just three matter particles which occupy the left-hand column: the up and down quarks, and the electron together with its neutrino. In moving one column to the right the basic pattern repeats, but this time it contains the charm and strange quarks and the muon (μ) and its neutrino; this column is called the second generation. Second generation particles are more massive than those in the first, and are less stable. The pattern is repeated one more time, making a third generation of even more massive and unstable particles, containing the top and bottom quarks, the tau lepton (τ), and its neutrino. In all, the standard model contains twelve particles of matter, governed by three forces that result from the exchange of four force-carrying particles: the photon (γ), the gluon (g), and the W and Z bosons.

In our story of matter, we have, in a sense, arrived at our destination: a truly Democritan view of the world in terms of its basic building blocks: fundamental particles of zero size. The knife that cuts a piece of matter up in ever smaller pieces can, as far as we can see, cut no more. If the fundamental particles are really dimensionless points, interacting via the forces of nature, we arrive at a remarkable picture in which matter is *empty*, and it is the four interactions that give it shape and form. But if history is anything to go by, we should not be confident that we have all the answers. On many occasions, scientists have believed that they have finally reached the ultimate picture, only to find that nature is subtler than they had thought and that underneath there is a deeper reality.

Even though the Standard Model is currently the best-tested and most comprehensive theory of the basic components of matter, there are still plenty of mysteries and glaring omissions. We have already noted the absence of gravity from the picture. If you let go of an apple, it falls in the Earth's gravitational field. Matter therefore *does* interact with gravity, yet there is nothing in the Standard Model to explain this. We know that the gravitational field supports oscillations, because gravitational waves have been detected from violent events in the cosmos. The predicted force-carrier of the gravitational field is the *graviton*, but this particle has so far not been detected. Also, we do not know why there are twelve matter particles, grouped into three generations. Nor do we understand why quarks and leptons are different, and why fermions and bosons differ. In the 1970s, a theory, called *supersymmetry*, or SUSY for short, was put forward that intimately connects matter and force by proposing that every particle in the Standard Model has a supersymmetric partner, or superpartner. For every fermion, there is a SUSY boson and vice versa, effectively doubling the number of particles. Although supersymmetry explains theoretically the distinction between matter and force particles, the issue is that if supersymmetric particles do exist, they should already have been observed in the LHC; so far, there has been no sign of them.

There are further problems. The masses of the fundamental particles are believed to come from the way that the quarks and leptons and the W and Z particles interact with the Higgs field. We do not understand why these particles interact in such a way to give them their different masses. When it comes to the origins of mass it is clear that there is still much to discover. In addition, there is evidence that the amount of matter in the universe is significantly greater than can be accounted for in the standard model. This is the so-called *dark matter* and *dark energy*, which will be discussed in Chapter 9.

The origin of mass

One of the most basic questions about ordinary matter is: where does *its* mass come from? This is a different question from where do the masses of the fundamental particles come from, which we have just looked at. We know that 99.9 per cent of the mass of atoms resides in their nuclei. Nucleons are made out of quarks, which interact via the strong colour forces through the gluon fields. Quarks and other particles derive *their* mass via interactions with the Higgs field. We can ask: how much mass do the three quarks contribute individually to the mass of a proton?

A proton has a rest mass of 938.28 MeV/c^2. An up-quark is estimated to have a mass of about 2.3 MeV/c^2, and a down-quark a mass of about 4.8 MeV/c^2. For the three quarks the total is 9.4 MeV/c^2. But this is only about 1 per cent of the mass of the proton! Where is the remaining 99 per cent of the mass? The answer comes from applying Einstein's formula: $m = E/c^2$, which tells us that wherever there is energy, there is mass. It turns out that the remaining 99 per cent of the mass of a proton or a neutron is in the mass equivalent of the energy of the gluon colour fields and the kinetic energy of the quarks. This is the origin of most of the mass of ordinary matter—it is pure energy.

In this chapter we have looked at the intensive picture of matter—peeling off its onionskin layers to reveal the smallest most fundamental particles of which it is made. Only a handful of elementary particles make up the world: quarks, leptons, and the force particles, which appear in the Standard Model of Particle Physics. These particles are the field quanta of a few quantum fields. The elementary particles get their masses by interacting with the all-pervasive Higgs field, but the dominant source of the mass of ordinary matter comes from the energy of the quark and gluon fields inside nucleons.

The Standard Model is a towering achievement of science, but it is not complete. In Chapter 9, we will look at two missing and mysterious pieces of the jigsaw puzzle: dark matter and dark energy. But before that, let's set the scene for the very large scales of matter, by tracing the chemical history of the universe in the 13.8 billion years since the Big Bang.

Chapter 8
Where do the elements come from?

The medievalists believed the Earth to be surrounded by a 'crystal sphere' containing Aristotle's unearthly fifth element, quintessence. However, when 19th-century astronomers attached prisms to telescopes and split the light from the Sun and other stars into different colours, they saw spectra peppered with narrow atomic lines having wavelengths matching those emitted by atoms in laboratory experiments. The atoms in the stars are the same types as those found on Earth and that make up our bodies. Carl Sagan was famous for saying that we are 'star stuff'. He was right.

To understand where the chemical elements come from, we have to go back a fraction of a second after the birth of the universe in the 'Big Bang'. The powerful alliance of telescope and spectrometer which had first revealed chemical elements in stars, struck again in the 1920s when Edwin Hubble and Milton Humason used the then biggest telescope in the world, the 100-inch Hooker telescope at Mount Wilson in California, to measure the distances and the radial velocities of galaxies. Just as the siren on a moving vehicle drops in pitch as it speeds away from you, the wavelengths of the spectral lines from the stars in receding galaxies are displaced (or *redshifted* to longer wavelengths at the red end of the spectrum). The size of this redshift allows the radial recession velocities of the galaxies to be determined. Hubble's law tells us that the galaxies are moving away from us at speeds that increase in proportion to their distances. The universe is expanding, and if we imagine a movie of the expansion run backwards, we infer that there was a moment of creation, a time when all the galaxies came together at one point, 13.8 billion years ago. The Big Bang was infinitesimally small, infinitely hot, and infinitely dense.

The origin of matter

All the matter and energy in the universe erupted out of the Big Bang. At very early times of less than a microsecond, the energies were far greater than those that we can produce in our biggest particle accelerators, and we can only extrapolate our current theories back to these early times, an extrapolation which inevitably incurs uncertainties. The shortest length scale that has any meaning in physics is the so-called *Planck length* scale of 10^{-35} metres, a distance below which our ideas about gravity and spacetime are no longer valid, and quantum effects dominate. The time taken for light to travel one Planck length is called the *Planck time* (10^{-43} seconds). This is the smallest unit into which we can subdivide our unit of time and so we can say nothing about the physical conditions of the universe earlier than this time.

As the universe expanded, it cooled. This cooling is a natural consequence of the law of the conservation of energy. By the time it was a microsecond old, the universe had cooled to a temperature of 10^{13} K (or 1 GeV), which is equivalent to the mass-energy of a proton. It was therefore too hot for the quarks and gluons to bind together to make nucleons. The universe would have consisted of exotic quark–gluon plasma or *quark soup*, a searingly hot cauldron of quarks, gluons, and photons, too energetic to stick together. This is thought to have been the only time in the history of the universe when free quarks could have existed.

At this time, conditions were so extreme that matter and energy were freely interchangeable. The laws of physics do not distinguish between matter and antimatter, so if precisely equal amounts of it were created and annihilated with complete symmetry, we would have had a universe filled only with radiation, and no matter. Yet, 13.8 billion years later, we observe that there are about 10 billion photons for every proton or neutron in the universe. This puzzle is known as the *baryogenesis problem* and raises the question: where is all the antimatter? It is of course possible that there are entire galaxies made from antimatter. An antigalaxy would have to be 'cordoned off' from normal galaxies; otherwise galaxy–antigalaxy pairs would annihilate each other in enormous explosions of gamma rays. Such huge explosions have not been observed. We believe that the 10^{80} protons of matter that constitute the visible universe are protons and not antiprotons, and that they were made from quarks left over from the vast number of quark–antiquark annihilations that are inferred to have taken place at early times. The Russian physicist Andrei Sakharov suggested a solution to the baryogenesis problem: for every 10 billion antiquarks formed in the early universe, there were 10 billion and *one* quarks, leaving a net balance of matter over antimatter. This model requires a very small bias in the laws of physics to favour the formation of matter.

Where could such a bias have come from? The slight asymmetry in the laws could be related to the violation of a type of symmetry called *parity*, or P-symmetry, the inability of nature to distinguish between the world and its mirror image. This is included in the Standard Model, and relates to the weak nuclear interaction. In 1956 Chinese-American physicist Chien-Shiung Wu, at the suggestion of T. D. Lee and C. N. Yang, set up an experiment to measure the magnetic spin of radioactively beta-decaying cobalt-60 nuclei. By aligning the spinning nuclei in a magnetic field, she found that the nuclei shot more electrons out from their south poles than the north. This was a great surprise because it showed that there is an absolute difference between the two poles of a nucleus. The asymmetry is called the 'failure of conservation of parity'.

Another symmetry of the physical laws is *charge conjugation* symmetry, or C-symmetry. It is a transformation such that, if particles are switched with their antiparticles, the signs of all charges are changed. This is true for all the forces *except* for the weak interaction, for which C-symmetry is violated. Experiments have shown that a type of meson called a *kaon*, first seen in cosmic rays, decays into pairs of pions for which the combined CP symmetry can also be broken during the weak interaction. There is a small tendency for decaying kaons to produce more positrons than electrons, and these experiments showed that nature has an inherent 'handedness'.

The immense numbers of quark–antiquark annihilations in the early universe produced electromagnetic radiation that cooled down as the universe expanded. The wavelengths of this afterglow radiation were stretched out by the expansion of space itself, and it is now observable as the 2.7 K *cosmic microwave background* (CMB) radiation that bathes the universe (Figure 30).

30. The full sky map of the cosmic microwave background radiation, the afterglow of the Big Bang, showing the infant universe. The mottled appearance is due to temperature fluctuations, and these provide information on the large-scale distribution of matter at early times.

As the universe expanded, temperatures fell below 10^{13} K (1 GeV), and it became energetically favourable for quarks to condense into protons and neutrons. By the time that the universe was three minutes old, the temperature had dropped to 10^9 K (100 keV), and the first nuclear reactions could take place to synthesize the light elements of matter. Nuclei of deuterium, helium-3, helium-4, and lithium-7 were formed in a complex of reactions called *Big Bang nucleosynthesis*. During this epoch, it was hot enough for nuclear reactions to fuse a significant proportion of the protons into helium but there was not enough time to synthesize many heavier nuclei. The primordial raw material for further processing in stars was therefore 'locked in' at this time, resulting in a composition of 75 per cent hydrogen, 25 per cent helium, and traces of other light elements.

It was however still too hot for neutral atoms to exist, and the universe continued to expand as plasma, in which the photons scattered copiously from free electrons. To an embedded imaginary observer, the universe would have looked like a dense fog. By an age of 380,000 years the temperature had fallen to some 3,000 K, when electrons could combine with nuclei to form the first stable atoms, at which point the fog cleared ('recombination'). From then on, the photons were no longer strongly coupled to matter and could travel freely through space. The photons that emerged from the edge of the fog bank, the '*surface of last scattering*', are the oldest we can now detect in our telescopes as the cosmic microwave background radiation. These ancient photons carry with them information about the physical conditions in the universe at this stage of its evolution, and show that the first large-scale aggregations of matter had already started to form. The CMB temperature fluctuations shown in Figure 30 are very small, representing 1 part in 100,000 of the average 2.7 K temperature that we observe. In the subsequent expansion, the density fluctuations continued to grow and concentrate primordial matter into the stars, galaxies, clusters, and superclusters of galaxies.

Furnaces of matter

By about 500 million years, the density fluctuations in the gas had grown into distinct clouds that became detached from the general expansion of the universe, and these started to collapse in on themselves under their self-gravity. The clouds fragmented into smaller clumps during a starless period called the *Dark Age*. The gas accreted on the cores of what would become the first stars, and was compressed and heated, just as the air in a bicycle pump gets hot when it is squeezed. At first the hot atoms radiated their heat into space, but then as the density increased, the energy was trapped inside, creating a thermal pressure to counter the inward pull of gravity.

At this stage a nascent star has a stable mass and size, and is called a *protostar*. As the core of the protostar reaches the temperature at which fusion reactions between protons can occur (around 15 million degrees), the hydrogen starts to burn liberating energy, which is transported out to the surface by photons and radiated into space. The temperature of a star increases strongly with its mass. The first stars were massive and hot and shone with an intense ultraviolet light, which ionized the surrounding gas. This epoch is called the era of *reionization*. The ionization of interstellar gas around newly born stars continues today in active star forming regions such as the Orion nebula in the Milky Way.

With the appearance of the first stars, the second phase of the production of the chemical elements, *stellar nucleosynthesis*, could then get going. To appreciate how stellar nucleosynthesis works, we need first to understand a basic concept relating to the stability of nuclei. Consider the stability of different-sized nuclei. In a small nucleus with only a few nucleons, a large fraction of the particles lies on the surface, where they enjoy the short-range 'Velcro-like' strong force binding them to their neighbouring nucleons with which they are in close contact. However, since the nucleons on the surface feel the binding forces only towards the centre of the nucleus, they can be 'evaporated' relatively easily from it. This is the reason why very small nuclei are not the most stable. At the other extreme, a large nucleus with many protons produces a strong repulsive electrostatic force, which tries to break the nucleus apart. The heaviest elements tend to be spontaneously radioactive. A large nucleus such as uranium, which is on the verge of instability, can push out an alpha particle (the origin of alpha radioactivity), removing two units of positive charge, and so becomes more stable.

The stability of nuclei can usefully be pictured on a diagram of the binding energy per nucleon (Figure 31). The binding energy curve has a 'U' shape, with a valley where the most stable intermediate mass nuclei (such as iron with 56 protons and 30 neutrons) are found. Less energy is needed to remove a nucleon from an element high up in the diagram than one lower down in

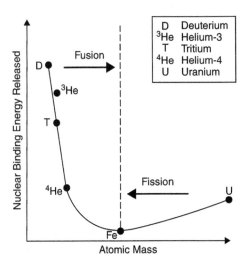

31. **The curve of binding energy per nucleon for various nuclei. Energy-producing nuclear reactions move in the direction of the most stable nucleus, iron, at the bottom of the curve.**

the valley. The nuclei high up in the diagram are therefore less stable than the ones lower down. The light elements: deuterium, tritium (the hydrogen isotope with one proton and two neutrons), helium-3, helium-4, etc. are at a high level on the left, and the heavy unstable nuclei such as uranium are above the valley on the right. Energy-producing nuclear reactions can proceed in two possible directions, always moving towards the more stable elements near the valley floor. Fusion reactions combine light nuclei into heavier ones, and fission reactions involve the splitting of heavy nuclei into smaller fragments.

The nuclear reactions in stars are thermonuclear fusion reactions, where light elements fuse together to make heavier ones and liberate energy. There are two main parts to a star, the hot *core*, where fusion takes place, and the surrounding low-density envelope, which forms a blanket and is the part of the star that you can see. When two nuclei fuse in the core of a star, they must approach each other closely enough to be gripped by the strong nuclear force. But, being positively charged, they have first to overcome their mutual electrostatic repulsion, and this means they must have high energies and high temperatures. The easiest reaction is between two protons. Two protons combine to produce deuterium, an initial step in the so-called p-p chain of fusion reactions that leads to the production of helium. The p-p chain produces the energy that powers stars of comparatively low mass like the Sun.

The heavier elements are built up in a 'building block' approach, in which the stable units of helium nuclei (alpha particles) fuse together to make bigger nuclei. When the theory of this process was first being worked out, there appeared to be a snag. It seemed that element building could never get over the first step, the fusion of two alpha particles. Two alpha particles produce an unstable nucleus, beryllium-8, which decays rapidly away before a third alpha particle can join it. There is no stable nucleus of mass 8 in nature. In 1953, English astronomer Fred Hoyle predicted that the nucleus of the next element in the series, carbon-12, must have what is called an excited state *resonance*. The concept of resonance is familiar to anyone who has seen a wine glass shattered by an opera singer. At its resonant frequency, the glass absorbs energy by sympathetic oscillations that grow in amplitude until the material fractures. In stars, the resonance of the carbon-12 nucleus has the effect of greatly enhancing the probability that carbon-12 will form before the beryllium-8 can decay. Hoyle urged experimentalists to search for the resonance, which they duly discovered at the energy he had predicted. Carbon is created in stars at a temperature of 100 million degrees by three alpha particles fusing together into the carbon-12 excited state, the so-called '*triple alpha*' process.

For bigger nuclei to fuse together, higher temperatures are needed to overcome their larger electrostatic repulsions. For example, in stellar carbon burning, two carbon nuclei, each with six protons, must fuse, while for oxygen burning the nuclei have eight protons. The core temperature of a star of around 20 solar masses can reach several billion degrees, which is high enough for elements up to nickel to be synthesized. The hot core of such a star is surrounded by an envelope of hydrogen and helium. Working inwards from the relatively cool outer surface to the hot core, there are nested onionskin-like shells of increasing temperature in which different nuclear reactions occur. The reactions proceed from helium burning in the cooler outer shell producing carbon and oxygen; these elements burn in the next hotter shell down producing neon and magnesium, then sulphur, and, in the high temperature core, silicon burning produces nickel. The fusion products of each nuclear reaction feed into the inner shells and the star evolves as it consumes its fuels.

In summary: the stars shine by liberating the fusion energy of the light elements, which sit high above the valley in the binding energy curve, to produce the middleweight elements, which sit lower down in the valley. By the time the star has consumed all of its easily available fusion energy fuel, it can get very little energy from burning silicon to make nickel, since both elements lie near the floor of the valley. The final silicon-burning stage is therefore rapid, lasting only a few days, and the nickel decays to form a stellar core of iron. The iron group of elements have the most tightly bound nuclei in the periodic table, and to make elements beyond iron requires an *input* of energy, which cannot occur by thermonuclear processes. As far as synthesizing the elements inside stars, iron is the end of the line. The question is: how are elements heavier than iron created?

The death of stars

When a star runs out of fuel, the thermal pressure that once kept it inflated vanishes, and the inexorable inward pull of gravity causes it to collapse. Stars have essentially three possible fates: a low mass star like the Sun will collapse into a *white-dwarf* star, a middle-range mass star can form a *neutron star*, or if the mass is high enough, matter is compressed into the most compact form possible in a *black hole*.

In a white dwarf, the gravitational collapse is halted by a new source of pressure: *electron degeneracy pressure*. This quantum mechanical pressure comes from the high energies of the electrons, when they are compressed into a small volume, in line with the uncertainty principle. The density of matter in a white dwarf is about a million times the density of water. There is however a limit to how much electron degeneracy pressure can be provided. If the mass of the collapsing star is larger than a critical mass called the *Chandrasekhar mass* (about 1.4 solar masses), the electron degeneracy pressure is insufficient to prevent further collapse.

When a 20 solar mass iron-cored star runs out of fuel, the Chandrasekhar limit is exceeded, and electron degeneracy pressure cannot support the star. The central pressure plummets, and the star can no longer fight against the inexorable pull of gravity. The iron core begins to collapse releasing a huge amount of gravitational potential energy, and implodes, reaching one-third of the speed of light in a second. The temperature rises dramatically and the iron nuclei disintegrate into their constituent nucleons. The electrons and protons merge via the weak interaction to form a neutron-rich form of matter. If more than a small fraction of its nucleons remained as protons, the electrical repulsion would be overwhelming.

Suddenly something extraordinary happens. Another new pressure source, the quantum degeneracy pressure of the *neutrons*, sets in and halts the collapse dead in its tracks. Unaware of what has just happened in the core, the rest of the star continues its headlong infall. When the outer layers of the star hit the now-stable core, they bounce back violently, sending out powerful shock waves. The star explodes as one of the most violent events that we know of in the universe, a *supernova*, flinging the outer layers of the star into space carrying their rich cargo of elements.

A supernova is a nuclear explosion with a yield that is a massive 10^{27} times bigger than a man-made H-bomb. The supernova shines with a luminosity a billion times that of the Sun, and for a few months can outshine its host galaxy. The expanding blast wave contains many solar masses of material, consisting of middleweight elements and dust grains, immersed in a flux of high-speed neutrons. The expelled nuclei rapidly absorb the neutrons, which are converted to protons via beta-decay, transmuting the nuclei into elements from lead up to uranium.

The neutron-rich stellar core is destined to become a *neutron star*, one the most exotic forms of matter of which we know. In effect it is a giant atomic nucleus held together by gravity, an object the size of an average city, weighing about twice as much as the Sun. A teaspoonful of neutron star matter weighs a billion tonnes, about the same as Mount Everest.

There is a symbiosis between stars and the gas clouds of the tenuous interstellar medium out of which they form. In their death throes, stars inject heavy elements into the medium to enrich the primordial gas, and the mixture gets recycled back to make new stars. The Sun and the solar system are 4.5 billion years old, and were made from matter from at least one generation of earlier stars, which had burned out billions of years before the Sun and our solar system formed. The elements of our familiar world, the carbon on which life is based, the oxygen that we breathe, and the iron in our cars all come from the stars.

There is roughly one supernova explosion in our galaxy each century. On 4 July AD 1054 Chinese astronomers recorded the position of a 'guest star' in the sky. Modern astronomers looked in the same direction, and discovered the Crab nebula (Figure 32). The filaments of gas in the nebula are rushing outwards into space,

32. The Crab Nebula is the remnant of a star that exploded in AD 1054.

and if their motion is backtracked, they must have come from a single point, the supernova explosion of 1054. This is what a supernova looks like after 1,000 years.

The alchemists' dream realized

The collapsed neutron star remnant of a supernova rotates rapidly, dragging round with it an immensely strong magnetic field. Such objects can be *pulsars*, objects that emit beams of electromagnetic radiation sweeping round, like the beams of lighthouses. Pulsars produce a sequence of precisely timed pulses, like clocks, as their beams sweep past our direction. The Crab nebula contains a pulsar that pulses thirty times a second.

The first *binary* pulsar was discovered in 1974, and consists of two neutron stars closely orbiting around their common centre of mass. This system is of tremendous interest because it allows general relativity to be tested in the strong gravitational fields and curved spacetime near highly collapsed objects. The two rotating masses distort space around them, which swirls around, generating ripples in spacetime, gravitational waves carrying away energy at lightspeed. The two stars are steadily losing energy, whirling ever faster around their common centre of mass as they come ever closer. The orbital period of the binary system has shortened by about a minute in the time since its discovery, as their rotational energy is converted into radiation. (The Earth orbiting the Sun also loses energy by gravitational radiation, but fortunately for us at an immeasurably small rate.) After a final rapid *'inspiral'* phase, the binary neutron stars will eventually merge in a cataclysmic merger that will wrench the fabric of spacetime violently and release a strong pulse of gravitational wave energy.

Gravitational waves were detected for the first time in 2015, from pairs of merging black holes. Astronomers had also been anticipating the detection of gravitational waves emitted when two neutron stars collide and merge. On 17 August 2017 their patience was finally rewarded when three gravitational wave observatories, two in the USA (LIGO) and one near Pisa in Italy (VIRGO), detected a burst of gravitational radiation from the merger of two neutron stars (an event called GW170817). The timing of the cosmic signal at the three sites enabled the physicists to triangulate its position with enough precision to correlate the event with a bright gamma ray flash recorded 1.7 seconds later by NASA's Fermi space telescope. The observation of GW170817 triggered an alert that galvanized 100 teams of astronomers to search for an optical counterpart. This they found in a galaxy 130 million light years away. Over the ensuing weeks, observatories detected electromagnetic radiation at X-ray, ultraviolet, optical, infrared, and radio wavelengths.

Merging neutron stars do not just produce gravitational waves; they also spew out hot dense chunks of matter at speeds of up to half the speed of light. The expanding cloud of debris forms a fireball in which the protons and neutrons combine rapidly to form heavy nuclei. These nuclei capture more neutrons, making then unstable and highly radioactive. The neutrons in the nuclei are converted into protons via slower beta-decay processes, and release energy which lights up the fireball. Spreading out into space, the fireball contains a rich cocktail of some of the heaviest elements in the periodic table. The aftermath of the collision is known as a *kilonova*, a bright transient event, less luminous than a supernova, but 10 million times more than the Sun.

Observations of the spectral lines emitted by these elements, using the *Very Large Telescope* in Chile, revealed signatures of heavy rare earth elements (lanthanides). This provided evidence that heavy elements in the periodic table (elements from niobium to uranium) were created in the merger. It has been estimated that GW170817 has ejected a few times the mass of the Earth in gold and platinum into space, exceeding even the wildest dreams of the alchemists!

The 2017 kilonova was the first ever detection of two colliding neutron stars, and was the first astronomical event in which gravitational and electromagnetic waves were observed together. It heralded the birth of a new astronomy: *multimessenger astronomy*. The near-simultaneous arrival of gravitational and electromagnetic pulses from an event which happened 130 million years ago is itself remarkable and means that propagation speeds of gravitational and electromagnetic waves differ by no more than 1 part in 10^{15}, in line with predictions of Einstein's theory of relativity.

In summary, there are three key processes by which the chemical elements have formed in nature. The light elements were synthesized in the Big Bang, the middleweight elements were (and still are being) forged inside stars, and the heavyweight elements were (and still are being) produced in violent stellar explosions and cataclysmic events. The elements from which you and I are made were produced in the lives and deaths of stars that existed billions of years before the solar system formed. The hydrogen atoms in our bodies go back to the Big Bang itself.

In Chapter 9 we will look at two mysterious and dominant forms of matter in the universe: dark matter and dark energy.

Chapter 9
Dark matter and dark energy

When we look into deep space with our telescopes, we see a cosmos filled with a hundred billion galaxies. Each galaxy contains around a trillion stars, and many galaxies are similar to our own, the spinning disc of the Milky Way. On the largest length scales, from the size of asteroids up to that of the visible universe, matter is dominated by a single force: gravity.

Gravity is the weakest of the four forces of nature. It is the force that keeps our feet firmly planted on the ground, and reaches out into space to guide the planets in their orbits around the Sun, and grips the trillions of stars as they swirl around their home galaxies. The motion of the planets is described extremely accurately by Newton's law of gravity. In discovering the law, Newton's brilliant idea was to imagine that the force on a falling apple and the Moon are really the same; both bodies move in Earth's gravity, only for the Moon the force is diminished in strength by its greater distance from the Earth. He imagined a cannon on a high mountain, firing shots towards the horizon at ever-higher speeds, so that they land further and further away. Eventually one goes fast enough to circle the Earth at a constant height and so becomes a satellite.

Newton's thought experiment tells us that the velocity of a satellite, its height, and the mass of the body around which it orbits are connected. If, for example, the mass of the central body is large, the speed of the satellite must also be large to remain in orbit and maintain the balance between its outward centrifugal force and the increased inward pull of gravity. Measuring of the size of the orbits of stars in galaxies and their orbital speeds is the basis of the method that astronomers have used to measure the masses of galaxies.

The stars in a disc galaxy orbit around the centre with speeds that depend on how far out they are. If the bulk of the gravitating mass of a galaxy is assumed to be concentrated in the middle, where there is often a prominent bulge containing a high density of stars, the rotational velocities of the stars in the outer disc should decrease the further out you go. This decrease is seen for the planets of the solar system, where the Sun's gravity weakens with distance. In the 1970s, American astronomer Vera Rubin measured the rotational velocities of the discs of nearby galaxies and found that, contrary to expectation, the rotational speeds did not decrease with distance, but remained constant. The galaxies were rotating too quickly for the visible matter they contain. When she worked out the masses of the galaxies, she discovered that there was about five times as much mass as could be accounted for by the combined mass of the stars and gas.

The Swiss astronomer Fritz Zwicky had reached a similar conclusion in the 1930s when measuring the motions of galaxies that swarm together in clusters. In the 1,000-galaxy-strong Coma cluster he discovered that the galaxies in the outer parts of the cluster were moving much faster than expected given the amount of visible mass in the cluster. This suggested that the cluster was being bound together by unseen gravitational mass, which he dubbed '*dunkle Materie*', or *dark matter*. The inference was that most of the mass in galaxies and clusters of galaxies is invisible matter, which we now believe to be distributed in large haloes enveloping the visible parts of the galaxies. This so-called 'missing mass' reveals itself by its gravitational interaction with normal matter, but does not give off or reflect light in any way.

The method of measuring the masses of galaxies used by Rubin and Zwicky relied on Newton's law of gravitation. But, as we saw in Chapter 4, Einstein in his general theory of relativity showed that mass curves space and curved space not only influences the motion of massive bodies, but it can also bend light. This light-bending aspect of gravity provides a powerful way to measure the mass of large bodies.

The light that was once emitted by the atoms in the stars of the most distant galaxies billions of years ago is only now reaching the Earth. In its long journey, the light has had to travel through vast distances of billions of light years. If, somewhere along its route, a photon from one of these very distant galaxies happens to pass near to a massive body and encounters the curved space that surrounds it, the light is bent and ends up on a different trajectory. This is similar to what happens when light rays are bent or refracted when they pass from air into a transparent medium such as water or a glass lens, of the type you might find in a magnifying glass. The exciting idea that comes out of this is that it is possible to use this light-bending effect, called *gravitational lensing*, to measure how much mass is present in space and how it is distributed. In a gravitational lens, the only thing that matters is the curvature of space; light can't tell the difference if the curvature is caused by dark matter or by the stars and other matter in a galaxy.

To help visualize how gravitational lensing works, consider Figure 33, which shows the tiled mosaic on the bottom of a swimming pool, viewed looking vertically downwards through the water. Ripples on the surface of the water behave like refracting lenses, and distort the image of the background tiles, which represent the distant background galaxies. We can think of the ripples of the surface of the water as being analogous to the

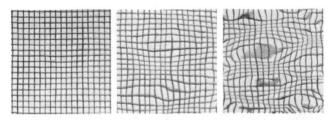

33. Illustration of the gravitational lensing effect, with water ripples in a swimming pool. Snapshots of mosaic tile pattern on the bottom of a swimming pool when the water surface is: (left) still; (centre) weakly rippled, and (right) strongly rippled.

curvature of space associated with the concentrations of intervening mass in the universe. There are three cases. When the water surface is still, there is no distortion and the tiling pattern appears to be regular; this corresponds to an empty universe containing no intervening matter. If the water surface is gently rippled, the grid pattern appears weakly distorted, a case that corresponds to *weak lensing*; this would be analogous to the presence of small concentrations of mass for gravitational lensing. Larger ripples create more extreme distortions, which break up the pattern into multiple images. This corresponds to *strong lensing*, and the presence of larger masses.

Strong gravitational lensing can be produced by the curving of space around large mass concentrations, such as clusters of galaxies, and can stretch out the images of background galaxies into long luminous arcs (Figure 34). Weak gravitational lensing is more commonly observed and, as the name suggests, is a less dramatic form which changes the shapes of background galaxies in more subtle ways. In this case, information on the mass of the lens can still be derived through the statistical analysis of the shape distortions of the very many background galaxies whose lines of sight pass close to it. In the swimming pool analogy, the amount of distortion to the tiling pattern provides information on the size of the ripples in the water.

34. The *Smiley Face*; the gravitational lensing of distant galaxies (curved arcs) by the mass of a nearer cluster of galaxies. Two large galaxies in the nearer cluster form the 'eyes' of the face.

Gravitational lensing has recently joined the armoury of modern astronomical techniques that will inform on the quantity and distribution of matter in the universe, particularly through observations that will be made with the coming generation of dedicated new telescopes. Gravitational lensing has been observed by the Hubble Space Telescope, but only in small patches of sky of a few square degrees. To assess the amount of dark matter there is in the universe, it will be necessary to survey a much larger part of the sky. One telescope that has been designed to do this is the European Space Agency's visible and infrared space telescope, EUCLID, due for launch in 2020.

What is dark matter?

The simple answer is that nobody knows. At present we know more about what dark matter *isn't* than what it is. There are two main ideas for what it might be. One is that it is simply ordinary matter, but in a form that absorbs or emits little or no light. This possibility comes under the broad heading of *Massive Compact Halo Objects* (MACHOs), which include failed low-mass stars (or brown dwarfs), Jupiter-sized planets, white dwarfs, and compact objects such as neutron stars. The problem with most of these possibilities is that, being made of normal matter, MACHOs will absorb and emit electromagnetic radiation—namely they ought to 'glow' at various wavelengths. Objects that glow in this way appear to be ruled out by observations.

The second and leading possibility for dark matter is that it is an exotic type of subatomic particle, known as a *Weakly Interacting Massive Particle*, or WIMP. WIMPs could have been made in the Big Bang along with quarks and radiation. A WIMP would need to be heavy (having a mass of between one and 1,000 proton masses), be stable for at least the age of the universe, and interact at most only weakly with the other particles in the Standard Model of Particle Physics. There is no particle currently in the Standard Model with these properties. Other options for WIMPs include the possibility that nature is supersymmetric. The lightest superpartner to a Standard Model particle, the hypothetical *neutralino*, is a possible candidate dark matter particle. Neutrinos are now believed to have very small masses of a millionth of the mass of the electron, and have therefore been considered as possible WIMP candidates. Neutrinos come under the category of 'hot' dark matter, which in this context means that they move at speeds close to the speed of light.

The fact that dark matter constitutes roughly six times the amount of ordinary matter that we know of means that it must have played a pivotal role in controlling the growth of the large-scale structures of galaxies and clusters of galaxies in the early universe. On very large length scales, the universe is smooth and uniform, but it is 'lumpy' on the scales of galaxies and clusters of galaxies. The role of dark matter in the formation of these structures has been investigated using computer simulations, with the objective of explaining the structures we now observe in the local universe. The simulations have been able to reproduce successfully the properties of the observed structures, but only on the basis that the gravitational pull is supplied by 'cold' dark matter (CDM). The 'cold' in CDM means that the velocity of the dark matter is assumed to be very much smaller than the speed of light, and it cannot cool by emitting photons, since it is dark. Hot dark matter tends to smooth out the small-scale structure too much. This, and the smallness of their masses, appears to rule out the candidacy of neutrinos as WIMPs.

Dark matter particles that may inhabit the halo of the Milky Way would be expected to stream continually though the disc of the galaxy and therefore should pass through the solar system. If they do, there is the possibility of direct detection when they strike the Earth. If there is a weak non-gravitational interaction between dark and normal matter particles, dark matter particles might be revealed in rare collisions with the nuclei of ordinary matter. A possible fingerprint would be the production of a photon from a dark matter particle interacting with a heavy nucleus. An experiment set up to search for such events is the *Large Underground Xenon* experiment (LUX), which is a large tank of liquid xenon, surrounded by banks of highly sensitive photomultiplier detectors. LUX is sited 1.5 km underground in the Homestake gold mine in South Dakota, deep enough to block out spurious particles. The chance of a WIMP strike on a xenon nucleus is expected to be very low and there have been no significant detections to date.

Dark energy

After publishing his general theory of relativity in 1916 Einstein went on to apply his equations to the cosmos. At the time the universe was believed to be static. There was a problem: if you fill a model universe with a number of masses and let it go, the masses fall together under their gravitational attraction—the universe couldn't remain static. Einstein tried to rectify this by adding a repulsive term to his equations, called the *cosmological constant*, usually denoted by the Greek symbol Λ. We can think of the Λ-term as a kind of antigravity force warping spacetime, working in opposition to gravity and pushing objects apart.

However, within a few years of Einstein's proposed cosmological model, Hubble had announced his discovery that the universe was expanding and not static. Einstein's original equation, without the Λ-term, could have explained an expanding universe. On hearing of Hubble's discovery, Einstein famously discarded his cosmological constant saying that it was the 'biggest blunder of my life'. But, as we will see, there is fresh evidence to suggest Einstein's reaction may have been premature.

Will the universe continue to expand forever, or will it one day start to contract, eventually ending in a big 'crunch'? The fate of the universe depends on a competition between the kinetic energy of the expansion of the Big Bang and the gravity of all the matter in the universe trying to pull everything together. The recession velocities of the galaxies that led to the discovery of the Hubble law are related to the kinetic energy of expansion and opposing it is the gravitational pull of matter. If the matter density is too small, its gravity is too weak to stop the universe expanding forever. If the density is too large, then eventually the expansion will stop, the universe will contract, and there will be a big crunch. But if the density has a finely tuned critical value, the universe will keep on expanding forever, and space will have a 'flat' geometry. The critical density is very small, equivalent to about five hydrogen atoms per cubic metre (for comparison, there are about 10^{25} hydrogen atoms in a glass of water). The evidence seems to indicate that the average density of the universe is close to the critical one.

In the 1990s, two groups of astronomers were trying to measure the geometry of the universe by pinning down the Hubble expansion as tightly as possible over extremely large distance scales. They were observing a very bright class of stellar explosions in distant galaxies, called *type 1a supernovae*. Type 1a supernovae are important because they generate nominally the same amount of luminous energy, and so can be used as 'standard candles' for distance measurements. (By measuring the *apparent* brightness of a known standard candle, its distance can be inferred from the inverse square law.)

The astronomers found that the most distant supernovae are consistently much fainter than had been expected. The surprising conclusion was that the space through which the light had travelled had expanded more than expected, and that the supernovae are further away than had previously been thought. This implied that the expansion of the universe is *accelerating*, which is not what is expected from a universe filled with gravitationally attracting masses. It is as if you throw a ball up into the air, and just as it is starting to slow down, it accelerates away from you and keeps on going. That is how surprising the result was.

An acceleration in the expansion of the universe is however exactly what Einstein's Λ-term can provide. The cosmic repulsion it engenders has become known as *dark energy*, a mysterious unknown form of energy that fills space. The dark energy *density* is constant, which means that as the universe expands and creates more volume, the total amount of dark energy it contains increases in step with the expansion. This in itself is an extraordinary notion. We believe that dark energy is spread smoothly across the universe and that its mass-energy density is small. Within the volume of the Earth for example the mass equivalent of dark energy is one millionth of a gram. The puzzle is why it is so small. We know that empty space has a latent energy density called vacuum energy that is related to the virtual particles that constantly flit in and out of existence on the quantum scale. However, the quantum vacuum energy density falls short of the inferred dark energy density by an enormous factor of 10^{120}, and so the true nature of dark energy remains a great mystery. In the very early universe, the effects of dark energy would have been masked by the then much higher energy densities of matter and radiation. But as the universe expanded, the presence of the dark energy component has become more marked and it is only in the past six billion years that the rate of the expansion of the universe has become significantly affected.

Bringing it all together

All the astronomical observations have been brought together in the standard model of cosmology, the Λ-CDM model, which is founded on Einstein's general relativity and includes the cosmological constant and a component of cold dark matter. The model accounts for the primordial abundances of the light elements in the hot Big Bang, the angular scales of the initial density fluctuations of the cosmic microwave background radiation, the large-scale structure of the distribution of galaxies, and the acceleration of the expansion of the universe. One of the strongest constraints of the model is the spectrum of fluctuations in the cosmic microwave background (Figure 30). The Λ-CDM model fits these observations and is consistent with a universe with an age of 13.8 billion years, a flat space geometry with an average density close to the critical one, and a tightly constrained mix of mass-energy components of matter. The mass-energy of the universe is made up of: 70 per cent dark energy, 25 per cent cold dark matter, and 5 per cent normal matter (the familiar atoms, quarks, gluons, and leptons). The contribution of the energy density of the cosmic microwave background radiation photons and neutrinos is small.

The stark reality that emerges is that the ordinary matter of the atoms and molecules of our bodies, and those of all living creatures, the matter that is studied in biology, chemistry, materials science, and engineering and in much of astrophysics, constitutes *less than a twentieth* of the matter we believe to exist in the universe. Normal matter appears therefore to be just an 'impurity' in the matter that is really 'out there'. The striking and humbling fact is that we do not know what the bulk of the matter in the universe is.

To round off this story of matter, we might hark back to Feynman's remark, quoted at the end of Chapter 3—that the most important single fact to pass on to the next generations is that all things are made of atoms. Since 1970, when he wrote those words, the great advances that have been made in astronomy have uncovered the unforeseen and apparently dominant forms of matter: dark matter and dark energy. A great challenge for future scientists is to find out what these mysterious forms of matter really are.

Further reading

Peter Atkins, *Galileo's Finger* (OUP, 2003).

Jim Baggott, *Mass* (OUP, 2017).

Stephen Blundell, VSI *Superconductivity* (OUP, 2009).

Brian Cathcart, *The Fly in the Cathedral* (Penguin, 2004).

Frank Close, VSI *Particle Physics* (OUP, 2012).

Richard Feynman, *QED* (Princeton University Press, 1988).

John Gribbin, *Einstein's Master Work* (Icon Books, 2015).

John Polkinghorne, VSI *Quantum Theory* (OUP, 2002) (this book describes quantum entanglement, which is not covered in Chapter 5).

Martin Rees, *Just Six Numbers* (Weidenfeld & Nicolson, 1999).

Carlo Rovelli, *Reality is not what it seems* (Allen Lane, 2016).

Russell Stannard, VSI *Relativity* (OUP, 2008).

Paul Strathern, *Mendeleev's Dream* (Penguin, 2001).

Stephen Weinberg, *The First Three Minutes* (Basic Books, 1993).

Stephen Weinberg, *To Explain The World* (Allen Lane, 2015).

Frank Wilczek, *The Lightness of Being* (Allen Lane, 2009).

Index

Matter

CHAOS
A Very Short Introduction
Leonard Smith

Our growing understanding of Chaos Theory is having fascinating applications in the real world - from technology to global warming, politics, human behaviour, and even gambling on the stock market. Leonard Smith shows that we all have an intuitive understanding of chaotic systems. He uses accessible maths and physics (replacing complex equations with simple examples like pendulums, railway lines, and tossing coins) to explain the theory, and points to numerous examples in philosophy and literature (Edgar Allen Poe, Chang-Tzu, Arthur Conan Doyle) that illuminate the problems. The beauty of fractal patterns and their relation to chaos, as well as the history of chaos, and its uses in the real world and implications for the philosophy of science are all discussed in this *Very Short Introduction*.

> '...Chaos...will give you the clearest (but not too painful idea) of the maths involved... There's a lot packed into this little book, and for such a technical exploration it's surprisingly readable and enjoyable - I really wanted to keep turning the pages. Smith also has some excellent words of wisdom about common misunderstandings of chaos theory...'

popularscience.co.uk

EPIDEMIOLOGY
A Very Short Introduction
Rodolfo Saracci

Epidemiology has had an impact on many areas of medicine; and lung cancer, to the origin and spread of new epidemics. and lung cancer, to the origin and spread of new epidemics. However, it is often poorly understood, largely due to misrepresentations in the media. In this *Very Short Introduction* Rodolfo Saracci dispels some of the myths surrounding the study of epidemiology. He provides a general explanation of the principles behind clinical trials, and explains the nature of basic statistics concerning disease. He also looks at the ethical and political issues related to obtaining and using information concerning patients, and trials involving placebos.

FORENSIC PSYCHOLOGY
A Very Short Introduction
David Canter

Lie detection, offender profiling, jury selection, insanity in the law, predicting the risk of re-offending, the minds of serial killers and many other topics that fill news and fiction are all aspects of the rapidly developing area of scientific psychology broadly known as Forensic Psychology. *Forensic Psychology: A Very Short Introduction* discusses all the aspects of psychology that are relevant to the legal and criminal process as a whole. It includes explanations of criminal behaviour and criminality, including the role of mental disorder in crime, and discusses how forensic psychology contributes to helping investigate the crime and catching the perpetrators.

www.oup.com/vsi

GALAXIES
A Very Short Introduction
John Gribbin

Galaxies are the building blocks of the Universe: standing like islands in space, each is made up of many hundreds of millions of stars in which the chemical elements are made, around which planets form, and where on at least one of those planets intelligent life has emerged. In this *Very Short Introduction*, renowned science writer John Gribbin describes the extraordinary things that astronomers are learning about galaxies, and explains how this can shed light on the origins and structure of the Universe.

GLOBAL WARMING
A Very Short Introduction
Mark Maslin

Global warming is arguably the most critical and controversial
issue facing the world in the twenty-first century. This *Very
Short Introduction* provides a concise and accessible explanation
of the key topics in the debate: looking at the predicted impact
of climate change, exploring the political controversies of recent
years, and explaining the proposed solutions. Fully updated
for 2008, Mark Maslin's compelling account brings the reader
right up to date, describing recent developments from US policy
to the UK Climate Change Bill, and where we now stand with
the Kyoto Protocol. He also includes a chapter on local solutions,
reflecting the now widely held view that, to mitigate any
impending disaster, governments as well as individuals must
to act together.

www.oup.com/vsi

NOTHING
A Very Short Introduction
Frank Close

What is 'nothing'? What remains when you take all the matter away? Can empty space - a void - exist? This *Very Short Introduction* explores the science and history of the elusive void: from Aristotle's theories to black holes and quantum particles, and why the latest discoveries about the vacuum tell us extraordinary things about the cosmos. Frank Close tells the story of how scientists have explored the elusive void, and the rich discoveries that they have made there. He takes the reader on a lively and accessible history through ancient ideas and cultural superstitions to the frontiers of current research.

'An accessible and entertaining read for layperson and scientist alike.'

Physics World

PLANETS
A Very Short Introduction
David A. Rothery

This *Very Short Introduction* looks deep into space and describes the worlds that make up our Solar System: terrestrial planets, giant planets, dwarf planets and various other objects such as satellites (moons), asteroids and Trans-Neptunian objects. It considers how our knowledge has advanced over the centuries, and how it has expanded at a growing rate in recent years. David A. Rothery gives an overview of the origin, nature, and evolution of our Solar System, including the controversial issues of what qualifies as a planet, and what conditions are required for a planetary body to be habitable by life. He looks at rocky planets and the Moon, giant planets and their satellites, and how the surfaces have been sculpted by geology, weather, and impacts.

"The writing style is exceptionally clear and pricise"

Astronomy Now

SOCIAL MEDIA
Very Short Introduction

Join our community
www.oup.com/vsi

- Join us online at the official Very Short Introductions **Facebook** page.
- Access the thoughts and musings of our authors with our online **blog**.
- Sign up for our monthly **e-newsletter** to receive information on all new titles publishing that month.
- Browse the full range of Very Short Introductions online.
- Read **extracts** from the Introductions for free.
- Visit our library of **Reading Guides**. These guides, written by our expert authors will help you to question again, why you think what you think.
- If you are a teacher or lecturer you can order inspection copies quickly and simply via our website.

ONLINE CATALOGUE
A Very Short Introduction

Our online catalogue is designed to make it easy to find your ideal Very Short Introduction. View the entire collection by subject area, watch author videos, read sample chapters, and download reading guides.